The H&R Book of Perfume

The H&R Book of Perfume

H&R EDITION

Understanding Fragrance. Origin, History, Development, Meaning.

Johnson · London

Authors

Julia Müller, Hamburg, West Germany

With the professional advice of the following
Haarmann & Reimer employees: Rüdiger Hall,
Dr. Ullrich Harder, Dr. Rudolf Hopp, Dieter Klemme,
Dr. Jürgen Nienhaus, Egon Oelkers, Klaus Peters,
Dr. Wolfgang Pilz, Karl-Heinz Schmid, Dr. Wolfgang Sturm,
Peter Wörner

Dr. Hans Bräuer, Medical-Service, Munich, West Germany

Joachim Mensing, Psychologist and Sociologist,
Research Institute for Applied Esthetics,
Freiburg University, West Germany
Christa Beck, Cosmetician, Freiburg, West Germany

Concept
Lothar F. Kümper, Karl-Heinz Schmid, Holzminden,
West Germany

American Editing:
Helga Büttner, Pete Peabody, Peabody Corners, Inc.

Design/Graphics
Fred Niemüller, Ronnenberg/Benthe, West Germany

Coordination, Production
Bernhard Kott, Siegfried Rogas, Holzminden, West Germany

Lithography
Hilpert, Essen, West Germany
Jahn-Repro, Bielefeld, West Germany

Printing
Frohne Druck, Bad Salzuflen, West Germany

English edition published by
Johnson Publications Limited
130 Wigmore Street, London W1H OAT

Printed in Germany
ISBN 3-87261-049-X

Contents

Fragrance Guide Feminine Notes

Introduction to the classification of fragrances according to families. Color illustrations of approximately four hundred original bottles of perfume on the international market, classified according to fragrance families, each product with fragrance description including top, middle and base note. Information on the major ingredients which determine the character of each perfume. Information on the producer of the perfume.

144 pages with 430 color photographs and
1 multi-colored illustration.
Dimensions: 23.6 x 30.3 cm, Hard cover.

Fragrance Guide Masculine Notes

There is at present in preparation a "Fragrance Guide Masculine Notes" which follows the pattern set by the "Fragrance Guide Feminine Notes". The work will contain color pictures of approximately four hundred men's fragrances on the international market, with evaluations and odor-classifications.

Approximately 140 pages, with approximately 400 color photographs, and 1 multi-colored illustration.
Dimensions: 23.6 x 30.3 cm, Hard cover.

Foreword

The mystical world of fragrances has always projected a unique fascination. The precious scents and their exquisite containers have been charming, ever-magical attendants to culture since antiquity. It is astonishing that, so far as we know, only a limited sector within the broad spectrum of the fragrance world has been made public.

With the H&R PERFUME BOOK we have made it our concern to create, for the first time, a complete encyclopedia of perfumery, for the interested amateur as well as for all those whose work brings them into contact with perfumes. It is not our purpose to portray perfume as either mysterious or incomprehensible, but to provide a glimpse behind the scenes of an industry and a look over the shoulders of the perfumers at work.

From the cultivation and extraction of the raw materials to the psychological meaning of fragrances, this volume will give honest and objective answers to the many questions that keep being asked about fragrances, over and over again.

In addition to this Volume I and as significant supplements to it, I would like to recommend the recently published Volume II, the H&R FRAGRANCE GUIDE TO THE FEMININE NOTES and the soon-to-be-published Volume III, the H&R FRAGRANCE GUIDE TO THE MASCULINE NOTES. Each of these Fragrance Guides contains 144 pages devoted to the 400 best-known perfumes on the international market, illustrated in color with odor descriptions and classification according to type.

I would like to thank the firm Haarmann & Reimer of Holzminden and its people, whose knowledge, professional advice and support have been crucial to the preparation of this work.

Wolfgang Glöss
Publisher

The History of Fragrance – as Old as the Culture of Mankind

"How fair is thy love, my sister, my spouse! How much better is thy love than wine! And the smell of thine ointments than all spices! Thy lips, O my spouse, drop as the honeycomb: honey and milk are under thy tongue; and the smell of thy garments is like the smell of Lebanon. A garden inclosed is my sister, my spouse; a spring shut up, a fountain sealed. Thy plants are an orchard of pomegranates, with pleasant fruits; camphire with spikenard, spikenard and saffron; calamus and cinnamon, with all trees of frankincense; myrrh and aloes, with all the chief spices: . . ."
(from The Song of Solomon)

The history of fragrance is as old as mankind itself. Fragrance is the essence of luxury and refinement. Even before people began to decorate themselves, their utensils and their surroundings, they were trying to please each other and their gods through the use of fragrances. In the beginning, they simply used flowers, herbs and resins. But soon they discovered that the resins and balms gave off their fragrances more readily and more strongly under the influence of heat. "Per fumum" (Latin: "By smoke"), became the name of one of the most precious of cultural luxuries.

Per fumum suggests that fragrances originally were used only as gifts to the gods. To make your gods merciful, you sacrificed the most precious things you had. And fragrant substances gradually replaced animal sacrifices as the world became more civilized.

The people of this world have been taking comfort and joy from pleasing fragrances for a very long time. The world knew of fragrant substances and their preparation at least four thousand years before the birth of Christ. The techniques of pressing, boiling, drying, powderizing, maceration in fat, and even a simple form of distillation are ancient crafts.

There was a great need for fragrant substances in antiquity, because the humans who populated the world in those days, from China to Egypt, from Persia to Rome, perfumed themselves and their surroundings with a lavishness that would appear to us grotesque. Those who could afford it took baths in perfumed water. Not only was the body scented, but also the hair, the clothing, the bed–including the bedmate, the favorite slave and the horse. Also the tiles of the houses and the temples, the sides of the tents, the sedan-chairs, and even the sails of the ships were perfumed.

Documentation of these fragrance-orgies is to be found in the laws that finally ended them. The ancient rulers did not appreciate the constant drain of precious foreign currency into the fragrance-producing lands. So edicts were published in Athens and Rome which prohibited, subject to heavy fines, the importation of foreign fragrance substances. And along with the publishing of the edicts, the perfumers were expelled from the cities.

In most cases, however, dealing with fragrances and their production was an honorable

profession, often practised by priests but mostly by doctors and other educated men. And since perfumery was part of medical science, fragrance waters were for internal as well as external use.

In the Assyrian Books of the Herbs, a collection of cuneiform-inscribed tablets from the second millenium B.C., there appear not only recipes for established medicinal herb compositions but also for fragrant ointments and perfumes, among them a perfumed drink meant to cure bad breath.

Fortunately, no educated man in antiquity, whether doctor or philosopher, considered himself to be above dealing with perfumery. So, thanks to the Greek chronicler Herodotus (around 485–424 B.C.), who travelled throughout the entire known world at that time, we have reliable knowledge regarding the fragrance habits of the Egyptians and their Near-Eastern neighbors, and even more importantly, of the extraction processes for incense and myrrh.

While the Occident, in cultural and scientific matters at least, slept deeply on bearskins (the farther north the deeper), the Near- and Far-Eastern cultures were experiencing their zenith–especially China and Arabia. Islamic Arabia was known for producing the finest mathematicians, astronomers, doctors and other scientific specialists. Furthermore, the people loved beautiful and sensual enjoyments, especially fragrances of all kinds.

The Arabs not only had an understanding and appreciation of the pleasures of the senses, but also a surprisingly advanced knowledge of hygiene and medicine. They produced herbal and other plant or animal elixirs for both therapeutic and cosmetic purposes. The great doctor and scientist Avicenna (980–1073) was the most famous of the Arabian doctors and philosophers who handed their knowledge over to succeeding generations. But there were many others.
Into the dark middle ages of the Occident, the Arabs not only exported empiric knowledge, but also two scientific dreams:

The dream of "The Philosopher's Stone" and the dream of "Aliksir," the Life Elixir. "Al kimija" was the Arabian word for "The Science of the Moist Things." The Greeks called it "Chemeia." From this point on, with true alchemist passion, European scientists–and also quacks and charlatans–attempted to make gold from non-pre-

cious material. And they also tried to make an elixir which would insure never-ending youth and health.

The alchemists found neither their philosopher's stone nor their elixir. But in all their trial-and-error, they developed technologies that are still in use today. They not only re-invented the processes for the manufacture of gunpowder and porcelain for the European area, but also made strides in the area of processes for distillation, a skill that came to be of tremendous value to the perfume industry. In those days, the Occident could only dream of the treasures of the Orient including the fabled Scents of Araby. Certainly, some had experienced them at first hand. But the deliveries of fresh supplies, because of the long, difficult and dangerous trade routes, were so infrequent and irregular as to be almost non-existent. Amber and musk, for the finely-engraved little spheres that people wore on necklaces and belts, were worth their weight in gold. The use of fragrances and exotic spices was a luxury reserved only for the very rich.

In palaces, castles and cloisters, there was much talk of the Holy Land, and of expeditions in the name of God to rescue Christ's grave from the hands of the infidels. In reality, the battles were looting expeditions under the blessing of the Church. After the confusion of the mass migration and the subsequent wars that were fought over succession, titles, family honor and borders, the Europe of the Middle Ages was impoverished and unstable but virile and venturesome.

In 1202, Venice conquered Constantinople and as a result, held the power over the Mediterranean sea routes to the East. With that began the ascendancy and the wealth of Venice, and also the flourishing of the perfume industry in Europe. From their crusades, the Crusaders and their followers brought many exotic specialties home: delicate silks, precious embroideries, artistically woven carpets, aromatic spices, fragrant ointments and essences. Europe began to require more and more of these luxuries, and Venice organized the importation.

The trading of perfume ingredients was made easy by the early scientific work of the Arabs. Thanks to the invention of distillation, the fragrant substances could be shipped more cheaply, as extracts. Instead of bulky bales of dried plants and voluminous containers filled with easily-perishable balm oils; instead of barks, roots and wood, there were now simply tiny vials with concentrated essences making their way into the unbelieving but lucrative Occident.

Meanwhile, the search for the philosopher's stone

and the elixir of life continued. In Latin, the scientific language of the Occident, they were called "Lapis Philosophorum" and "Aqua Vitae" respectively. In alchemist parlors and scientist chambers, in cloister cells and the laboratories of doctors and pharmacists, the stoves glowed and the retorts bubbled, above all for the making of gold and the distilling of the water of life. The gold-making was unsuccessful, but the distillation made progress. After all, the distillation of wine did lead to a form of Aqua Vitae: Concentrated Alcohol! Even Albertus Magnus, one of the greatest scientists of the 13th Century, was involved in the production of alcohol, the "spirituous waters," to which were attributed numerous healing powers for rotting teeth and lame legs, pox and cholera, depression and madness, and, most importantly, for the preservation of youth.

Actually, alcohol proved to be a universally usable miracle. As early as the 15th Century,

spirits of wine was being used not only for drinking but also for preserving, and for the extraction of medicinal herbs, spice herbs and fragrant plants. It was at this time that Oleum Mirabile, an alcoholic extract of rosemary and resins, was produced. Oleum Mirable was taken internally for medicinal purposes. As time went by, many modifications were made in its composition. One was that its resin extract was left out. The remaining rosemary distillate, now diluted with alcohol, then began its successful journey around the world as "Hungarian Water." It was the forerunner of eau-de-cologne.

While Antiquity, and the Italy and France of the Middle Ages, had acknowledged the concept of the perfumer–a specialist in a specific field–the production of fragrant waters in Germany was firmly in the hands of the clergy. The cloisters were the basic centres of education, and the aromatic waters which were brewed there were used mainly as medicines. Monks and nuns, by no means unworldly, studied and copied scientific literature in their libraries, cultivated herbs and fragrance-bearing as well as other useful plants; and in these earliest of medical complexes, they formulated and experimented with various medicinal compounds. They also brewed beer and distilled other monastic beverages, not for therapeutic purposes alone. The "Herb Garden", the recipe collection of the Benedictine abbess Hildegard von Bingen (1098–1179), includes the description of a Scented Lotion, very famous at the time.

The cloisters provided the European market with alcohol and herb extracts for medicinal and pharmaceutical processing. Yet Perfumery as such did not make much progress, despite all the Christian zeal for research.

A different story in Italy, however. In the Fifteenth Century, Venice was the number one entry point for goods from the Near- and Far-East. The residents of the island republic were immensely wealthy and generously enjoyed the imported Oriental and Byzantine luxuries. The other harbor cities lived equally well. It was said that the perfume and spice needs of the entire world could easily have been filled from Naples alone.

In the flourishing cities and city-states, where the old aristocratic families competed with the up-and-coming merchant and banker families in opulence and splendour fragrances were always part of the display: fragrant lotions and balms; perfumed soaps; sweet, fragrant herbs; flower-scented powders; fancy make-ups; frankincense; bathing oils; spicy dishes and aromatic drinks–of which some were so poisoned that the most prominent people were forced to employ tasters. There was a persistent rumor abroad that murder could be committed with poisoned gloves. The rest of Europe looked on with a mixture of admiration, jealousy and disgust at such heights of worldly luxury and depths of infamy, greed and unscrupulousness. At the centre of all this sinful pomp lay the ducal and the papal branches of the Borgia family, especially the papal off-springs, Cesare and Lucrezia.

At the beginning of the Fifteenth Century, the Italians were by far the richest citizens of Europe. Everything the rest of Europe needed in the way of luxury goods, including perfume, cosmetics, and the raw materials for both, came through the Italian ports of entry. This was about to change.

In 1453, Venice lost Constantinople to the Turks; in 1489, the Portuguese sailed around Africa for the first time; in 1492 Columbus found the sea-route to the West Indian Islands. From then on, the seafaring nations brought home their Oriental –and later, South American treasures– in their own ships.

Around the time of these geographic discoveries, the Renaissance was beginning in Northern Europe. Along with the arts and sciences, perfumery was also developing. There is a legend that Catherine de Medici (1519–1589), the wife of King Henry II, brought modern perfumery to France. It is said that while on her bridal trip, in 1533, she stopped in Grasse and afterwards instructed the French court in the making of perfume. But this is doubtful. The bride was only a child of fourteen, and the arrogant Camarilla would not have taken advice from an upstart Italian Bourgeois anyway. The Medici were indeed rich, but still only a banker-family. . .

The French had been using the herbs of their homeland for medicines since the Middle Ages. They were cultivating fragrance plants and importing exotic essences through Marseilles, where soon a soap industry began to flourish. Yet the French courts were not considered very clean. Make-up, powders and fragrances were used more often than soap, and the people were as wasteful as their Italian counterparts.

It is a fact that a Florentine perfumer by the name of René, one of Catherine's attendants, opened the first Italian perfumery in Paris; and his fragrances, powders and beauty balms sold well. The perfume, however, was not made in Paris, but in Grasse.

This small town in southern France, at the foot of the Mediterranean Alps, was actually the center of the leather processing industry, full of busy tanneries. Plants from Provence supplied tannins and dyes, additional hides and chemicals were imported through nearby Marseilles, and the market was the famous Fair of Beaucaire. Tanners not only had to be good craftsmen, but also good chemists and salesmen. Besides wool

and linen, leather was one of man's most widely used materials, not only for boots and jerkins, purses and bags, belts and straps and gloves; but also for belts, aprons and many other articles in the clothing and accessories categories. The courtly habit of overall perfuming didn't stop at leather goods–just the opposite.

As no method had been invented for cleaning leather, it had to be heavily perfumed. The decoration and perfuming of gloves in particular was considered to be an especially fine art. In Grasse, the proximity of the tanning and growing perfume industries led to a combined guild of glovemakers and perfumers.

As a perfume city, Grasse had a number of geographic advantages. The Provence was a blessed herb garden, in which the imported plants from India, Persia and the Iberian Peninsula grew beautifully, especially citrus fruits but also carnations, tuberose and jasmine. Lavender grew wild there in profusion. In addition, there were these helpful factors: eager and hardworking craftsmen; technical knowledge, mainly from Moorish Spain; and fine import-export traffic connections

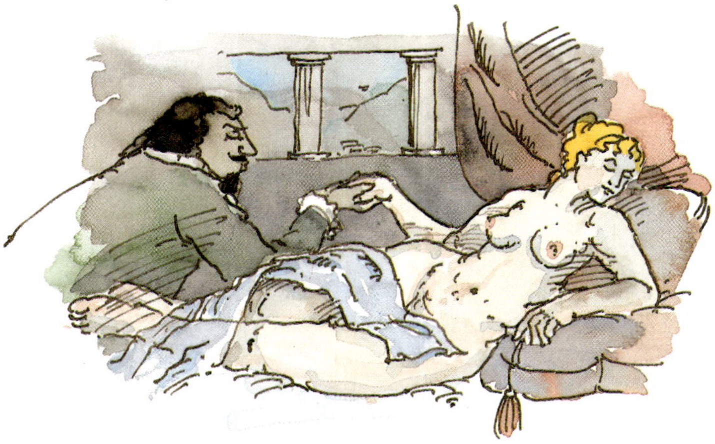

through Marseilles from the Mediterranean area and to the growing North-European centres. But most important: at the end of the Fifteenth Century, luxury, refined life-style, courtly culture, pomp and extravagance moved into France. And with it all, of course, perfumery in all its subtle refinements. There were perfume manufacturing facilities in France even before the Revolution. Perfume was then, and has remained a forte of the French people.

Methods of producing fragrances were improving, but the basic techniques were still distillation, expression and extraction in alcohol and fat. Extraction with volatile solvents represented a great step into modern times. It had been discovered that certain blossoms such as jasmine and tuberose did not yield essential oils through the steam distillation method. But the alternative, enfleurage, the method of extraction in fat which had been brought to a high point of efficiency in Grasse, required much labor and time. The growing chemical industry delivered its first solution to the problem in petroleum ether, a volatile solvent that could facilitate the extraction of essential flower oils. While the objective in Grasse was to increase the yield of natural fragrance substances and to refine the compositions, the frontiers of modern perfumery basically were to be found in German laboratories. The objectives in those laboratories were not actually fragrances, but the understanding of the structures of natural flavor and fragrance complexes, the syntheses of important ingredients; and the production of artificial fragrance substances. The last-named are bodies that are not found in nature and that consequently can give entirely new odor-effects. The elite of the classical perfumers looked askance upon such test-tube activities, but they could not stop the progress. Most women accepted the new, fresh and exciting fragrances enthusiastically. They were far different from the flowery scents and Oriental harem type perfumes they had been used to. The fact is, when the perfumers discreetly began to use the new fragrance substances, they had to admit that their creations became more individual, more striking, more brilliant.

The aldehydes, for example, which were introduced into some perfumes of the 1920's in dosages previously considered to be impossible, have given their name to an entire fragrance family.

Then and now, fragrance was and is a legitimate child of its time. It is one of the most joyous boons the earth has ever known.

Flacons, Pots and Jars – 5000 Years of Precious Vessels for Precious Fragrances

Perfumes there are as sweet as the oboe's sound, Green as the prairies, fresh as a child's caress,–and there are others, rich, corrupt, profound.
And of an infinite pervasiveness, like myrrh, or musk, or amber, that excite the ecstasies of sense, the soul's delight.
(From The Flowers of Evil–Beaudelaire)

Fine perfumery is an art, and it has inspired artists and artisans for thousands of years to create works of art as fitting containers. To do so, they have used a wide variety of materials: clay, stone, bronze, glass, porcelain, gems and precious metals.

The examples in the following illustrations can not come close to covering the entire subject. However, they may succeed in giving an impression of the variety and abundance of the materials; and, through the creations, demonstrate that the artists were experts in the use of them.

The fragrances which the containers once held have long disappeared for they were ephemeral, but through good fortune and favorable conditions, many of the containers still remain, both for our study and for re-living of the past.

Before glass, the most important raw materials available for this use were clay, metal and stone.

So we see in the first illustration from old Egyptian times, a typical ointment container made of alabaster. It is round and has no base, but rather a disc-like rim to allow it to be hung from a holder.

Next to it is a charming little perfume container with neither base nor rim–but it could be closed tight and carried around with one's personal effects.

From the classical Greek epoch, numerous ceramic receptacles in various shapes, carrying a multitude of different designs–such as these–still exist.

Notable on the *Lekythen* (ointment vials) are the deep black decorations on a light ceramic background, or the pale yellow drawings on a black background. These receptacles feature similar shapes: all with one handle, all with an offset neck and narrow shoulders. They vary, basically, in the diversity of their ornamentation. They were used in ancient funeral rites, stayed buried with the dead, and thus were beautifully preserved through the centuries.

Another typical Greek perfume and ointment container is the *Aryballos,* dating from approximately 600 to 500 B.C. It is a round-bodied oil flask with a narrow orifice, disc-shaped mouth and banded handle. This type of container had no standing base. It was carried on its owner's wrist, attached by a small strap. There exist famous, beautiful illustrations showing athletes with the Aryballos strapped around their wrists, cleaning themselves of the dust and the sweat of the fight with scented oils and scrapers. This was a form of Greek personal hygiene, dating from the days of the old Olympic Games.

Popular motifs on these small containers were the so-called Hoplites (bearers of arms, with shield, helmet, sword or spear) as well as the renowned "Owl of Minerva."

The slender-shaped *Alabastron,* standing on a platform, is another typical fragrance container from old Greek times. Next to it there is a so-called "Krater," a double-handled ceramic vessel with a wide mouth, used for mixing water and wine. This type of large ceramic-ware has been of great significance to research–its impressive, glossy decorations allow glimpses into the lives and activities of ancient Greece, including the customs pertaining to the use of cosmetics and perfumes in those times.

From the cultural epoch of the Etruscans, the predecessors of the Romans on the Apennine Peninsula, here are ceramic receptacles for fragrances and cosmetics. *Alabastron,* Aryballos and an ointment container with a large mouth and lid, typical of those employed in normal daily use. Next, two Etruscan Lekythen, with colored designs on a

black background, and a magnificant Aryballos, ovoid with a flat base, small mouth and disc-shaped orifice.
Pictured on it is a proud Etruscan lady, admiring herself in a hand-mirror, her features beautified by cosmetics and an artistic hairstyle. A pity that we cannot also experience her perfume; it would be of great interest to our noses. Certainly it was as much in tune with fashion as perfumes are for the elegant women of today in their endeavours to be attractive to themselves and to others. How little has changed in the last two and a half thousand years.

For five hundred years, the Roman Empire shaped the Occident in politics and culture. The invention of the glass-blower's pipe came in the First Roman Empire. It revolutionized container-manufacturing in Sidon, in Syria, at that time part of the empire. Glass-blowing opened up a completely new field of possibilities for the design and production of articles made of that old material, glass, and took over from the old, time-consuming sand-grain manufacturing method.

Our illustrations of Roman glass in abundant variety demonstrate the result of this revolution. All glasses were blown, mostly free-blown, but also at times with the help of pre-designed molds. Through this development, mass production became possible and prices tumbled. Almost 2000 years before, when the manufacture of glass-ware began, the material had been worth its weight in gold, and pieces of art containing it were of the highest value. (Example: Tutankhamon's death-mask.)

Besides the large group of "plain" perfume bottles, in all shapes and colors–also called *Balsamarien,* tear bottles or sprinklers–there are other pieces of artistic importance. For example, iridescent twin Balsamariens with handles; a small urn-shaped vase with dark handles; a well-known date-container; and two sprinklers adorned with glass-threaded ornaments.

The most impressive in this small collection of Roman glass–notable not so much for any special value as for its large number of shapes–is a beautiful dark blue iridescent container in the shape of an abstraction of a small bird, the beak serving as the spout. In this piece, the artist produced an outstanding effect with the simplest of means.

The iridescent effect of old glassware is produced by a natural decomposition process of the glass during centuries in the ground. But beware of fakes: newer glass can be rendered iridescent artificially. The effect is obtained by exposing the glass to metal vapors while it is still red-hot. The process is not foolproof, however. The trained eye may be able to recognize that some of the patina is missing.

This method has been used during different periods to produce the so-called "Iris and Luster"–extremely thin metal coatings on glass which possess a mother-of-pearl-like gleam. This effect is particularly fascinating on the Islamic glassware of the early Middle Ages, and on some of the magnificent perfume containers of the Art Nouveau Period. A master in this field was Lewis Comfort Tiffany of New York, whose creations in the Art Nouveau mode now command phenomenal prices.

In our stroll through the millenia, we've now reached the Arabic era. Due to the restriction in Islamic belief against the portrayal of living things, ornamentation was developed as the principal characteristic of Islamic art. Perfumes and perfume containers were "en vogue." Even the Koran mentions the use of musk, which was especially popular among the Arabs.

It was the Arabs who brought the distillation of rosewater to its prime in the 8th Century. During the 9th Century, they took a leading position in art and science.

In our illustrations we show–next to plain perfume containers made of clay and glass–the typical blue Arabian ceramic, and an imaginatively executed metal vial which probably contained rosewater. It is a well-known fact that the Arabs significantly improved distillation methods. Because of their work, the quality of distillates improved considerably, which in turn benefitted the quality of perfume raw materials.

Now a big jump forward in time. The Spanish art of perfume-production–always of great significance, with its roots in the Moorish culture–is represented by a magnificent perfume vial dating from the Eighteenth Century. It probably originated in Catalonia, where it bore the Catalan name *Almorratxa.* It is a receptacle with a rounded shape, standing on a single base, and was made mainly for the purpose of holding rose-water. Projecting from its shoulder-area, next to the neck, are four jet nozzles, which were used for sprinkling.

Now it is England's turn. In modern times, work in silver has been especially notable there. We see smelling bottles and snuff boxes in many different shapes, either fitted out with silver or made of silver. Noteworthy are the silver vials, among them a so-called "scent-fish" with moveable joints. Creative examples, such as the four vials in the mother-of-pearl casing, or the head with the helmet-like silver mounting, are expressions of an epoch in which a rich upper crust took a special delight in precious objects. Originating in the second half of the Nineteenth Century, a group of variegated perfume bottles, some with colored overtrails, all with metal fittings, partly of silver, partly of gold-plated silver. Fanciful, too, is the small ivory chest with perfume vials.

France, which during the reigns of her kings, produced marvelous perfume containers in unique colors, is represented here only by some perfumery brick-a-brack: little fragrance accessories worn on tiny chains by the elegant lady.

Germany, on the other hand, is fortunate enough to have been able to participate with some magnificient Meissen porcelain vials. They are in demand throughout the world as collectors' items, and accordingly are very expensive.

Finally, Art deco from France, the birth-place of this style–fanciful designs with attractive shapes. No claims to great artistic value, just good, honest industrial output. An unpretentious ending to this demonstration of the variety of style and form in which perfume containers can be produced.

At the beginning of this little essay, we referred to the fact that for around five thousand years, large numbers of artists, some very well-known, have been striving to create worthy containers for precious perfumes. And it continues, although in the course of technological development, particularly in the past century, the methods of production have mainly changed from works of art in the hands of meticulous artisans to industrial production. Top artists continue to make it their business today to design perfume containers. Even today, some of them still fashion unique and non-reproduceable containers from a variety of different materials, especially from free-blown glass.

So, even in the age of industrial production, works of art in the perfume vial category continue to be created and take their places in the galleries next to sculptures and paintings. Of these, the most interesting find their ways into museums.

Like the beautiful fragrances they contain, most perfume containers today are made by careful mass production methods. For compared to earlier times, the consumption of perfume has increased enormously. Now the whole populace buys perfume, not just priests and kings.

Mass production or not, one important point has never changed: the creation of a new fragrance and the design of a new container are still the work of artists. Whether it be a designer or a perfumer, his inspiration remains the cradle of the creation.

Egypt

Greece

Greece

Greece

Greece

Etruria

Etruria

Roman Empire

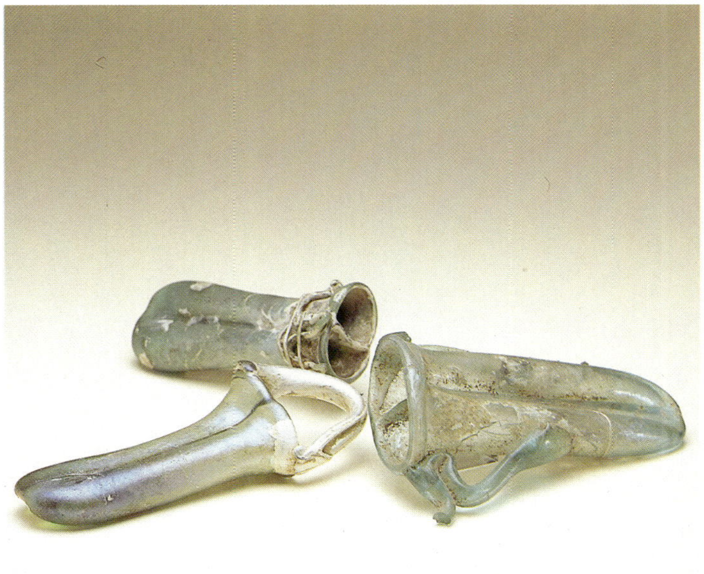

Roman Empire

Roman Empire

Roman Empire

Roman Empire

Arabia

Arabia

Spain

England

England

England

England

England

France

Germany

Germany

Art deco

Art deco

Art deco

Art deco

Changes in Fragrance Fashion over 8 Decades

Every age, every culture and every society has its own characteristic odor–and this does not only have to do with perfume. But since perfume is part of every woman's daily luxury, and since it is also artfully tailored to individual wishes, it would seem that every fragrance could be right at any time, provided that it has been cleverly presented. And yet, the fragrance favorites of any given epoch have always somehow reflected the spirit of their times. It is just as impossible for a perfume to be decreed by design as the right scent for a particular time as it is for the highs and lows of the skirt-lengths to be arbitrarily dictated by the makers of fashions. A perfume must simply fit its time exactly, as a natural happening, as jeans or tailored suits or mini and maxi skirts. A review of perfume evolution over the last eighty years indicates that it has always gone parallel to the social development. It appears that it was not necessary for the really successful trend-setting perfumes to come on the market at any particular time. Just the opposite. Some of the most outstanding creations were decades ahead of the times of their great popularity and success, while other heavily-promoted imitations lived only for a short time as minor fads.

1900 ——— 1910

Flower Fragrances lose their Innocence.

The gaiety and exuberance of La Belle Epoque did not move out of the salons and dance-palaces, boudoirs and séparées on the dot of the turn of the century. The moods of the time don't go by the calendar.

But the turn of the century, featuring the beginning of the use of electricity, marks the start of a scientific, technical era. Madame Curie discovers uranium radiation, the first zeppelin rises into the air, Caruso is recorded for the first time. Those times, when our grandfathers married our grandmothers, will remain in our memories as the good old days. Queen Victoria, who had influenced the previous era in all ways, died at a very old age (in 1901). However the clothing of the Bourgeoisie is still "Victorian": tight, dark trousers with high-buttoned jackets and stand-up collars for the men; corsets and dark, long and full skirts for the women–tufted, plushy, with trains trailing after. The scenes of a country idyll from Peter Rosegger's "When I Was Still a Farmboy" seem to describe a world in order.

The favorite flower fragrances of that time lost their innocence too. There is still a whiff of jasmine and violet from the last century, but single-florals have been increasingly broadened into bouquets. Around the turn of the century, Coty created his Rose Jacqueminot (1900), and, five years later, L'Origan, a powdery carnation fragrance. This direction developed later, via

Quelques Fleurs/Houbigant, into an abundance of floral fantasy compositions.

1910 — 1920

Like the Women of this decade, Perfumes became Emancipated.

When the "Titanic" went down, with most of her distinguished passengers, on her maiden voyage in 1912, it seemed to be a signal for the downfall of a seemingly well-ordered, stable, patriotic and tradition-conscious era. On the eve of World War I, the women's liberation movement was at its climax with the English Suffragettes; the main features of modern psychology had been established; and the dance on the top of the volcano was being accompanied by New Orleans Jazz. After the war, the world had radically changed. Working women, however subordinated, had become an accepted part of society. In 1918, German women won the right to vote; hobble-skirts and cart-wheel hats were only to be seen on old photographs. A few courageous ladies had their hair cut though up to that time it had only had contact with curling irons. At least the skirts had shortened to ankle-length.

The development of perfume had kept up with the times. Even though dazzling floral bouquets such as Quelques Fleurs/Houbigant and the orientally-inclined Narcisse Noir/Caron came on the market, it was the decade of the Chypre fragrances. In 1917, Coty created his legendary Chypre. It gave its name to a whole future fragrance family and from then on, together with Mitsouko/Guerlain, inspired many of the classics of subsequent decades.

1920 — 1930

With Bobbed Hair and Charleston to New Fantasy Notes.

In order to visualise the extravagances of the Wild Twenties, you should read "The Great Gatsby" by F. Scott Fitzgerald. The skirts slid obove the knee, the waist went down to the hip, the hair was short and the necklaces and cigarette holders were long. At the beginning of the decade, inflation had reached its climax. But a pleasure-seeking society, no longer so high-minded and ultra-respectable, danced on. And it was the Charleston that they danced.

Hits got to be hits through the phonograph now; it was a brand new development. Outwardly, life was gay in the midst of economic and social chaos. It was fertile ground for excess, decadent eccentricity, hectic amusement; but also unprecedented freedom of thought, satirical humor and courageous taste. Art Nouveau and Art Deco were and still are the only definitive styles to be generated by the new century.

At the same time, technical progress was uncritically admired. The fascination of machines, automobiles, airplanes and electricity; of dizzy speed and glittering capital cities overwhelmed the old and excited the young. Never before had there been an idolization of progress and youth such as this. "Modern" was the magic word.

Modern youth was especially concentrated in the cities. Heavily made-up girls à la garçonne and their escorts with slicked-down, brilliantined hair populated the dance palaces, beautifully described in the contemporary novel "Das Kunstseidene Mädchen" ("The Girl of Artificial Silk") by Irmgard Keun.

On the less spectacular side, the modern, self-confident working woman had grown up. For her, the young French couturier with a mind of her own, Coco Chanel, created the modern wardrobe: short, sporty dresses and above all suits that were elegant yet simple.

To go with her fashions, she had a perfume created, to which she gave the name Chanel N° 5. It was an aldehydic perfume, so called because of the aliphatic aldehydes at that time increasingly being used in the perfume industry. These synthetic fragrance substances permitted the composition of new, brilliant fantasy accords, which since the Twenties have been firmly established in perfumes with great character. Bois des Îles/Chanel and Arpège/Lanvin followed. Chanel and Lanvin had started a trend by going into the perfume business. Other couture houses such as Patou, Molyneux, Worth, Schiaparelli and Weil soon followed with perfumes of their own. All in all, the decade was one of the most creative in the history of perfumery.

1930 — 1940

Oriental Notes and the Comeback of the Classical Fragrances.

In "Congress Dances", the Germans cheered their favorite movie stars, Lilian Harvey and Willi Fritsch; but the dancing had come to an end. Little Man What Now?, asks Hans Fallada in his 1932 novel about unemployment. As quickly as the skirt hems had gone up, down they fell again, to a new length below the calf. As the skirts came down, the hair went up. A recrudescence of

prudery went as far as the “Gusset Edict” against immodest swim-suits.

As the book-burning began in Germany, and thousands of her intellectuals began to emigrate, the world’s number one excitement centred around the scandal of the decade: King Edward VIII had renounced his throne to marry the American divorcee Wallis Simpson.

The fragrance favorites of the time were the intensive, unmistakably Oriental notes: Tabu/ Dana and Shocking/Schiaparelli came on the market; Jicky/Guerlain, Narcisse Noir/Caron and Shalimar/Guerlain which were already in existence, found new devotees. Also, Cuir de Russie/Chanel, a leather note, fitted into the popular trend of the voluptuous and voluminous fragrances. At the same time, beautiful flowery perfumes were developed, in particular Joy/ Patou. Joy, at that time, was the most expensive perfume in the world.

1940 — 1950

Better Times – and Green Notes.

“Outside, Waiting at the Door” is a rough translation of the title of the German returning-home novel by Wolfgang Borchert. There were millions of refugees trying to establish themselves in the Germany that remained. Many owed their survival to the Care Packages from the U.S.A. There was a black market in which the currencies were nylon stockings, cigarettes and coffee.

The upswept hair-style rapidly took on. Instead of the head-scarves and dusty dresses of the women who cleaned up Germany’s bombed-out cities, the “New Look” was enthusiastically embraced in that country, as it was in so many others. It featured voluminous, swinging, medium-length clothes giving a feminine silhouette with natural waist-line. The permanent wave was ten years old, but it had a resurgence at the end of the Forties because of the popularity of masses of tiny curls. The Italian Vespa became the vehicle of youth. And youth wanted to dance bebop instead of the waltz and listen to jazz instead of Lili Marlene.

The recovery brought fresh winds and the new direction of the “green” fragrances such as Vent Vert/Balmain and ever more subtle chypre notes such as Ma Griffe/Carven and Miss Dior/Dior.

But the sensation was Femme/Rochas, an outstandingly feminine, mossy-fruity chypre fragrance.

1950 — 1960

Rock ’n’ Roll, Glamour, Jeans and Chypre.

While in the German film version of the American soap operas, the *Heimatfilme,* sweet farm girls and worthy, earnest foresters romanced, in the U.S.A., cult films were being produced, such as “East of Eden”, with James Dean portraying the young iconoclastic rebel. The world fell in love with Marilyn Monroe, the classic, sexy platinum blonde of “The Seven Year

Itch". While the older generation regaled itself with detailed accounts of the life of the Persian Empress (she gave the name to a whole new style of journalism–"Soraya Press"), the so-called "Skeptical generation" of sociologist Helmut Schelsky discussed "The Diary of Anne Frank." The generations also drifted apart in terms of fashion. Dior put the ladies into the H, the A and the Sack Silhouettes, while the young ones, in blue jeans, screamed for their beloved Elvis Presley in rock 'n' roll rhythms.

The preferred fragrances, too, became more aggressive. The green chypre notes, which had first become popular in the Forties, returned to fashion in animal-floral variations. Even though Bandit/Piquet and Miss Dior/Dior were the first representatives of this new trend, the big success came with later developments. Intimate/Revlon was the main American example and Cabochard/Grès the European. Now, logically, an Oriental trend was to be expected.

It came in 1952, with the name Youth Dew/Lauder, an outstanding success in the U.S.A., but Europe still remained relatively untouched by the Oriental trend.

1960 — 1970

"Let a Thousand Flowers Bloom" – New Floral and Aldehydic Notes.

The decade that began with orderliness and prosperity and seemed to end successfully, with the first human being landing on the moon, was comparable to the Twenties in terms of its inner strife. Youth and women's liberation were both significant, as they had been in the Twenties. Youth didn't trust anybody over thirty, and the pill took women great strides ahead in their emancipation. Their bodies, they said, belonged to them. They

still stalked around on stiletto heels, in tight skirts that extended to just below their knees; teased their hair tower-high; and painted Brigitte Bardot pouts on their lips. But the Flower Children had started on their psychedelic trips and The Beatles–"Yeah, Yeah, Yeah"–played their guitars and popularized the long male hairdos that became the symbol of protesting youth. Meanwhile, Mary Quant started cutting her skirts thigh-short, and anyone who felt she didn't want to try the mini-skirt could turn to pants-suits. Along with all of this came "demonstration fashion"–boots, jeans and parka. Ideal for demonstrating in the streets against the Establishment, the Vietnam War, and the usages of a thousand years.

Rebellious youth, with its leanings toward Eastern teachings and meditation techniques, amused itself with musk fragrances and incense, but in regular, main-stream perfumery, radiant light-floral fragrances were developing.

And so the basic ideas of Quelques Fleurs/Houbigant, L'Air du Temps/Ricci and Diorissimo/Dior were taken further to Fidji/Laroche and Norell/Revlon. While the enormously successful Fidji, a light, flowery fragrance, was tailored more to the European taste, Norell, with its stronger fixative, was in line with the American fragrance preference.

At the same time, as a parallel theme, the aldehydic notes were being developed. Building on the success of Chanel N° 5 and Arpège/Lanvin, Madame Rochas/Rochas and the equally elegant Calèche/Hermes were introduced. The former was characterized by its exclusive rose note, the latter by its incense component, in tune with the times, which put it in the neighborhood of the chypre notes.

Since the Fifties, two important perfume trends had prevailed in each decade. In the time that would come, the changes would come more rapidly.

1970 — 1980

The Scene Changes; the Carrousel of Fragrances keeps turning. . .

If you really tried to keep up with the fashion and the ideology of the Seventies, you could easily get out-of-breath. The Student-Rebellion Generation began to make peace with the system and climbed up into the teaching and managerial chairs. The consumerism that they had formerly despised had become a habit; and father, with his thinning hair and mother in her folklore dress looked on aghast as their offspring from the anti-authoritarian day-care centers developed into champagne-drinking, well-groomed Poppers. The whole scene split apart into a great number of groups which wanted to have nothing to do with either the older people or the other groups. The commune-people would have nothing to do with big-city chic, the anti-establishment leather-

jackets would have nothing to do with the Greenies. The lesbians would have nothing to do with the mothers in their bib-overalls and their nursing infants. The emancipated wanted nothing to do with the career-women, and the softies and the machos did not want to have anything to do with each other. But practically all could agree on "The Limits of Growth". The first energy crisis of 1973 was only the beginning of a recession which would become the leading political topic by the end of the decade. The lights of the discos that had sprung up like mushrooms were already starting to go out. What is preferred is what is allowed, the makers of fashion had always said. But in those days, virtually no fashion could be sold without individual exceptions–each group had to have its own uniform.

The abundance of the perfume developments of that time was as confusing and many-faceted as the social fragmentation. Starting from Cabochard, the chypre direction brought increasingly cool floral accents, at the beginning of the Seventies, into compositions such as Ellipse/Fath, Empreinte/Courrèges and Sikkim/Lancôme.

At the same time, the green notes were becoming more and more popular. One of the first great successes of the Seventies was Chanel N° 19, an original green, flowery composition with dominating orris accents.

Next: the elegant, sporty green Alliage/Lauder, which became an inspiration for many more creations. Also floral notes – never really out of the running – had much success since the end of the Sixties. Two essentially floral American perfumes must be noted: Estée/Lauder and Charlie/Revlon. Then a little later, came Chloé/Lagerfeld, the first in a line of tuberose perfumes, to which Jontue/Revlon, Belle de Jovan/Jovan and Enjoli/Ritz also belong. The aldehydic notes also made a successful comeback during this period, latching onto established successes.

Climat/Lancôme, Calandre/Rabanne, Chamade/Guerlain, Rive Gauche/St. Laurent and Farouche/Ricci were all presented as new developments in the richer, woody-powdery direction. Mystère/Rochas, is a new flowery aldehyde perfume with a dry amber background.

A rediscovery in the Seventies was the Oriental perfume category. Opium/St. Laurent is the pinnacle of a whole line of successes which had begun in the Fifties with Youth Dew/Lauder, and produced in the Seventies refined, but frankly erotic perfumes such as Vu/Lapidus, Expression/Fath, J'ai Osé/Laroche, Cinnabar/Lauder and Magie Noire/Lancôme.

Further green notes and floral notes followed, both in sparkling, young versions with radiant top-notes that have just a nuance of fruitiness: Amazone/Hermes and Janine D./4711 in the green direction. As starting points for the "white blossom" direction came Quartz/Molyneux, Anais Anais/Cacharel, and Metal/Rabanne.

The Eighties.

Elegance and Precious Fragrances.

Where the fashions of the current decade will lead us remains to be seen. Certainly the Eighties brought a turning point. Politically, due to the election of conservative governments. Socially, due to more consideration for virtues such as diligence and discipline. Simultaneously, an increasing consciousness of the environment developed. After the past euphoric, seemingly unlimited economic growth, a phase of keeping and caring began. The joiner took the place of the dropout. America called these people "yuppies" (young urban professionals).

This young, city-living, highly-qualified successful-person appears in a three-piece pinstripe suit in the male version and a tailored suit with silk blouse in the female. And now youthfulness–so far as it can be connected with success–lasts well into the fifth decade of your life ("Don't trust anyone under thirty.").

According to the internal logic of perfume development, the "white blossom perfumes" were confronted at the beginning of the Eighties with a new generation of Chypre-types: Exclusive new developments such as Chimère/Matchabelli, Espris/Factor, Senchal/Ritz, Andron/Jovan, Sculptura/Jovan, Le Temps d'Aimer/A. Delon, Gianni Versace/G. Versace, Daydreams/Maybelline, Missoni/Missoni, Volcan d'Amour/v. Fürstenberg, and Diva/Ungaro.

In parallel to these developments came fragrances which can be characterized as "floriental"

–complex floral fragrances with an Oriental touch. Oscar de la Renta/Stern marked the beginning of this trend in 1976. It succeeded in the Eighties with new products such as Halston Night/Halston, Vanderbilt/Vanderbilt, Raffinée/Houbigant and Balahé/Leonard. In the same period, a trend towards lighter aldehydic perfumes appeared in which–as in "white blossoms"–fresh, fruity floweriness is combined with a new kind of rose accent on a classical basis. A forerunner in this category was 1978's White Linen/Lauder. In addition there came in 1983, the violet-rosy Paris/St. Laurent, which featured a strong diffusion of floral sweetness.

The analysis of past trends has shown that fragrance preferences come in waves–from light fragrances to heavy ones, and back to light again. Even though the changes take place more rapidly now, the order has remained the same.

As a matter of fact, it seems to be impossible for one perfume type to remain predominant for more than a few years. Just as there is no longer a homogenous society to dictate taste and fashion, it also seems to be true that there is no dominant type of fragrance in the affections of most of the world's women. Furthermore, perfume is used more freely nowadays, as part of day-to-day living. Depending on her mood, the occasion, the time of year or the time of day, the same woman uses different perfumes, just as she wears different clothing for different occasions. This palette of perfumes is called her "fragrance wardrobe". It actually says more about the individuality of the lady than it does about any kind of a perfume trend.

1980 ? 1990

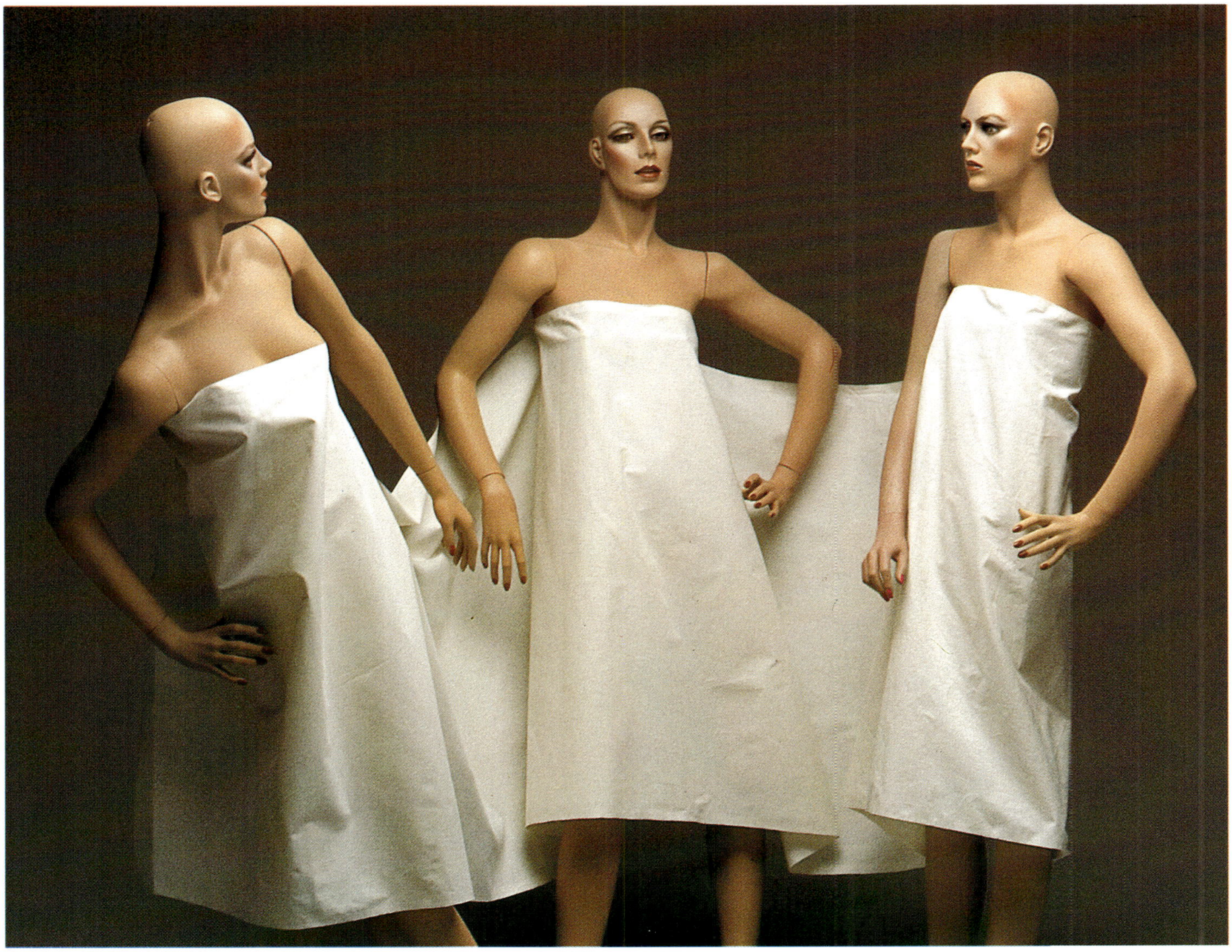

Fragrance in Various Consumer-Products. Its Function and Effects

"Perfume is the nourishment that stimulates my thinking," said Mohammed. Of course, the Prophet saw fragrances in a different light than the woman of today, but beautiful fragrances have traditionally been an important part of the sensuous Arab culture. Those who use perfume today are probably hardly aware of the stimulating–as well as calming and relaxing–effect of fragrances. Perfume is now much more an expression of the social personality of an individual. It helps people indicate to those whom they would like to please, who they are, what they are, or how they would like to be.

The lay-man thinks of all alcoholic fragrances as "perfume." The fragrance expert differentiates between them according to the degree to which they have been diluted. Technically, perfumes are mixtures of divers substances, some of vegetable or animal origin, others wholly or partly artificial. Dissolved in alcohol, these mixtures of various volatile fragrant substances release their scents into the air at normal temperatures.

Only the extrait–the mixture which contains the highest proportion of fragrance concentrate and the least possible alcohol–is called "perfume." Any mixture containing a larger percentage of alcohol is called Eau (water) although the mixture may be mainly alcohol, with little water present. Alcohol used in perfumery is never 100%; even "pure alcohol" still retains 4–6% water. The lower the quantity of fragrance

essence in a mixture, the lower the grade of the alcohol used.

The following table shows the concentrations for the different kinds of alcoholic perfumery. (Concentrations may vary from country to country and brand to brand.)

	Quantity of Fragrance Concentrate	Grade of Alcohol Used in Mixture
Perfume (or "extrait")	15–30 %	90–95 %
Parfum de Toilette or Eau de Parfum	8–15 %	80–90 %
Eau de Toilette	4– 8 %	about 80 %
Eau-de-Cologne	3– 5 %	about 70 %
Splash Cologne	1– 3 %	
Eau Fraiche	about 3 %	about 80 %

In addition to the fragrance solutions diluted with alcohol, there are also those which are diluted with oil–for example, "Youth Dew." And compact and cream perfumes are produced by mixing up to 25% fragrance oil with solids such as paraffin or other waxes.

Most women buy the Eau de Toilette or Cologne of their favorite perfume, basically because the "waters" are less expensive than the more heavily concentrated "perfumes." But also because they can thus afford to use the fragrance more liberally, and because the lighter solutions have a more refreshing effect. This is so much the custom that with some fragrance brands, there exists no "perfume" variant at all–for example Ô de Lancôme. Similarly, the Cologne Water, which gave Eau-de-Cologne its name, is also a perfumery product in its own right. It was invented in the year 1695 by the Italian Giovanni Paolo Feminis, who had emigrated from Italy to Germany and settled in Cologne. It has been sold since 1716 as Water of Cologne (Eau-de-Cologne) and is now a whole category of fragrance produced by many manufacturers. Eau-de-Cologne has always been a combination of neroli, bergamot, lemon, cedar and orange oils with a background of musk or amber. Formerly, it was used not only externally as a fragrance, but also taken as a medicine. But in the 18th Century, it became mandatory to make public the ingredients of medicines. Rather than divulge the secrets of their formulae, the manufacturers of the Eau-de-Colognes abandoned the medicine business and concentrated 100% on the fragrance area, which required no such ingredient declaration.

In 1806, "Eau-de-Cologne" was established as a brand name by Jean Marie Farina, a descendent of Femini who had settled in Cologne "opposite the Jülichplatz". It is said that one day a tourist bought a bottle of that famous fragrant water in a store in the city. When he had paid for it, he said to the storekeeper "Now that we have completed our transaction, please tell me frankly: are you the real Farina?" "No," the storekeeper answered, "the real Farina owns the store across the street." The tourist, impressed by such a display of honesty, went and bought another bottle in the store across the street. Back in the hotel, he told the hotel owner of his adventure, and ended his account with the revelation: ". . . so I've met an honest salesman in your town." The hotel owner smiled and said: "The store across the street was his affiliate and Farina–the real one or the false one, whatever the case may be–has found a way to sell two bottles instead of one."

You go Shopping with your Nose . . .

While the perfume in its beautiful little bottle has, for the most part, an aesthetic purpose only, there are hundreds of fragrances involved with our daily lives that have another job to do: they make products pleasant to use, and therefore more desirable to purchase–even though they have nothing at all to do with the efficacy of these products.

A further function for such fragrances has to do with product families–for example, a series of body-care products from the same manufacturer. The manufacturer gives them the same fragrance. This along with the package design, supports the family feeling. If the lady likes the soap, the chances are she'll buy the body lotion.

Fragrance extends through all levels of product-pricing. Not only is the most expensive luxury soap perfumed, just about every ordinary bar soap you'll find on any supermarket shelf is, too.

Practically all the products we use in the bathroom, kitchen and household in general are perfumed–the shampoo, the setting lotion, dishwashing detergent, household cleaner, wax-polish, insect-spray, hair dye, nail polish. In these cases, the fragrance is an adjunct with an introductory role, pointing to and supporting the function of the consumer-product. For example, the pine scent of a bath oil combined with a green color indicates Nature, naturalness and health. The lemon scent of a dishwashing detergent suggests appetizing freshness and "grease-cutting" properties. The fragrance of a product is also extremely important in influencing a first purchase. The consumer in this case has had no experience with the product, doesn't know whether it's worth purchasing. In these cases, the fragrance is often the factor that influences the sale the most. Tests made with a single cream perfumed with different fragrances have indicated that the more elegantly the product is perfumed, the more quality it is deemed to have.

Another important duty for product fragrances is the masking of very strong or unpleasant odors of products, such as heavy-duty household cleaners, depilatories, wave-setting lotions.

30 in a Single Day.

To demonstrate the ways in which fragrances surround us, let's consider a housewife on a typical day on which she does her household work, performs her toilet, and prepares to go out to a party in the evening. First comes the house-work. She washes the dishes with a liquid detergent, scrubs the pots with a cleaning pad. She uses a soap and a hand-lotion before she loads the washing machine with laundry, plus a water-softener, fabric-softener, laundry detergent; and starts the machine.

Now she begins the ritual of her daily personal care, with tooth paste, cleansing milk, face lotion. As her hair has become dull, she shampoos it and then applies a color-rinse, perhaps with built-in conditioner. Next a bath, made luxurious by a beautiful-smelling bath-oil. After the bath, she cleans the tub with a cleanser and then applies a body-lotion to herself. An eye-cream comes next, and then a face-cream. Then a deodorant spray. Before she dries her hair, she uses polish-remover on her nails and renews the varnish. If she has a blow-dryer, she uses a special setting-lotion. For her make-up, she applies a foundation first, then the eye-shadow, mascara, rouge. She finishes with a pat of powder and the lipstick. The hair is now dry. She combs it and applies a little hair-spray to keep it in place. She's ready to go to the party, but before she leaves, there's one last look in the mirror and the application of the perfume. On this one Saturday, for the purposes of our example, she's used thirty products, each of which contains its own scent. Not quite as far-fetched as it may seem. We've not come close to exhausting the field of perfuming consumer-products.

Every consumer-product makes its own, different demands on the perfumer's skill. Depending on the product, there can be aesthetic problems, or physical problems, or both. A soap perfume, for instance, has to be specially blended to avoid chemical reaction with the soap-base. A fragrance for household bleach must be robust enough to withstand chlorine, and a detergent perfume should be stable to alkali. Then again, the scent of a fabric-softener must be "substantive" enough to linger on in the laundry-cupboard, long after drying and ironing.

The bath-additive or oil is supposed to be relaxing, and further, it should give the whole bathroom a pleasant odor and stay on the skin for a while. The hair-spray must have a pleasant scent, but should not overpower or clash with her perfume. And, last but not least, every one of these perfumes is expected to remain stable after long storage and under differing temperatures.

It may seem from what is written here that modern society floats in an ocean of scent. But its fragrances are subtle and mild compared with the effluvia of French High Society in the 17th Century. They perfumed everything then: the hair, clothes, gloves, fans, belts, shoes, cushions, furniture, the wallpaper, and even the writing paper. All for a very good reason: in those days, taking a bath was not just some normal daily routine; it was a rare venture out into the dangerous and unknown. You did not do it until after you had consulted your doctor. And if he gave his permission, he would recommend that you take a whole day to recover from your bath. Perfume was lavishly used not so much for aesthetic reasons as for masking the unpleasant odors which were all around you.

The Sense of Smell: Its Function and Meaning for Emotion and Consciousness

Odorants – Signals of our Environment.

One of the most ancient inventions in the course of development of animals and humans is their olfactory sense, their sense of smell. Using this sense, the animal or human is able to obtain important information from is environment. Thus, a certain odor can be vital for the survival of an animal if it helps it find food or prey. Animals can also attract members of the opposite sex with the aid of their sense of smell, and mate in order to propagate their species. Among human beings, and even among higher animals, a feeling of desire or aversion can also be induced through the perception of odors. Our sense of smell can warn us against harmful substances and trigger appropriate protective reactions, e.g. flight. Inhalation of irritating gases, e.g. such strong vapors as mustard oil or vinegar, can even cause us to hold our breath.

How did the Sense of Smell Develop?

Life on Earth began in the water of the oceans, in an age in which only a portion of the mineral salts had been leached into the sea from the Earth's crust. Even today, the blood serum of human beings consists of a composition of salts that is similar to that of diluted sea water.

The first animal and vegetable life forms were protozoans, i.e. they consisted of only one cell. Later, as a transitional form between protistes and metazoans, or multicelled life forms, colonial species developed, such as sponges and corals, in which numerous protozoons live together in a community. For the protozoan in a colony of this type, the environment consists of its neighbors and of the ocean water that surrounds it. The protozoon is informed about the condition of its neighbors through direct contact. And this type of information transmission through direct contact from cell to cell still exists today in the nervous system of human beings (synapses). In human beings, the information that the protozoon was able to obtain from the ocean water that surrounded it corresponds to the manifold effects produced by the messenger substances – called hormones – that are brought to the body's organs and cells with the blood.

As life came ashore from the oceans, lakes and rivers of the world, it left its aqueous environment, from which it had previously been able to obtain information. Now, the environment was gaseous, and a biological trick was necessary in order to be able to receive the information blown in by the new environmental medium, air. It was necessary for the air to pass over a thin mucous layer, in which the substances carried in the air could be captured in order to then be tasted in the same manner as before. For an animal living out of the water, a direct perception of taste was only possible in the case of those substances that came in contact with its tongue; everything else could only be perceived in the form of an odorant.

As a means of remote perception, the sense of smell undoubtedly was originally even more important than vision, which only later developed its full capabilities. In the case of human beings, there is not much left of the capabilities once offered by the sense of smell, yet nevertheless, olfactory perceptions still play a significant role in our feelings and our response to the environment, even if we are frequently not fully aware of these influences. Odors can excite us or cheer us up, yet they can also have a repulsive, debilitating effect. In the form of sexual odorants, which are called pheromones, odors play an important role in relations between the sexes, just as do the messenger substances produced by our bodies, the hormones.

Odorants as Interpersonal Messages.

The sensation triggered by an odorant, an aroma chemical or a perfume is experienced by the entire person, as the signal triggered by the odorant in the nose and passed on via the nerves reaches important regions of the brain and, through complicated regulatory systems, initiates responses in the body's organs and fluids.

Legend 3
A three-dimensional representation of the olfactory epithelium. Located in the two halves of the upper nasal cavity are two approx. 3-4 cm^2 (1/2 sq. in.) large olfactory mucous membrane regions in which, packed between supporting cells, the actual olfactory nerve cells are located. (1) Mucous film, (2) Olfactory region, (3) Olfactory cilia with end segments, (4) Olfactory nerve cell, (5) Plump olfactory cell process, (6) Supporting cells, (7) Basal membrane, (8) Basal cells, (9) Schwann cell, surrounding the nerve fibers, (10) Olfactory nerve fibers, (11) Small olfactory nerve.

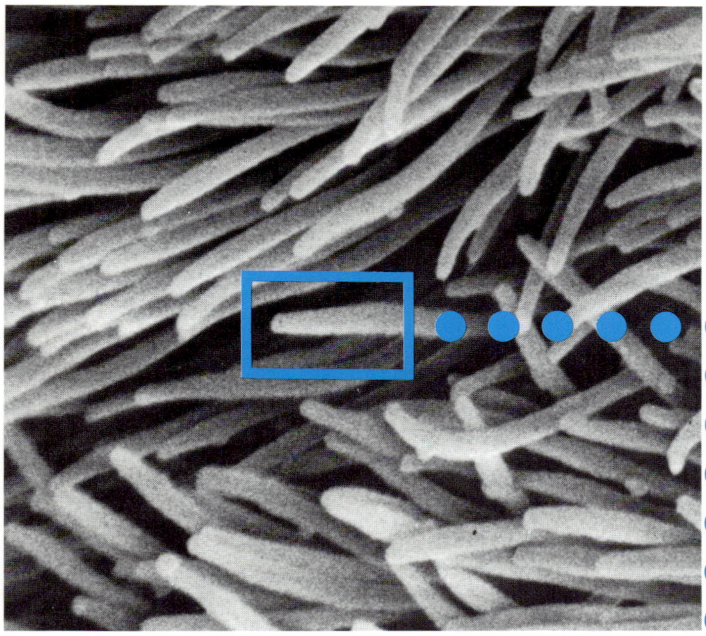

Legend 1
This electron microscope photograph shows the olfactory cilia and their slender tips. They come from the plunge processes of the olfactory cells and are blanketed by the mucous film above them, which has been rinsed away from the cilia here in order to provide a better view. Magnification: 10,000 power.

Legend 2
Terminal part of an olfactory cilium, embedded in the mucous film on the surface of the olfactory mucous membrane. Odor molecules have penetrated into the mucous film from the air and now occupy their matching receptors. The molecules are representative of certain basic odors. (1) Camphor, (2) Peppermint, (3) Musk, (4) Ethereal, (5) Floral.

We also Smell with our Brain.

Our sense of smell is located in two regions of our nasal mucous membranes, each about 3-4 cm^2 (1/2 sq.in.) in size, on the roof and the upper part of both walls of the nasal cavity, protruding from the center into the lateral regions of the nasal cavity in the form of thin plates of bone, covered by mucous membrane (Fig. 3). These 3-4 cm^2 (1/2 sq.in.) large regions do not uniformly consist of cells that secrete mucus, like the rest of the nasal mucous membrane, but also carry slender nerve cells, covered with a thin film of mucus, between equally slender, so-called supporting cells, that also produce their special, aqueous mucus as a secondary function (Fig. 1). The nerve cells send a plunge process into the mucous membrane, from which numerous cilia protrude. It is here that the complicated mechanism of smell begins. In simple terms, it can be described as follows. On the surface of these cilia, there are tiny depressions, whose chemical structure is such that the molecules of certain odorants fit into them exactly, like pegs into holes (Fig. 2). As soon as an odor molecule that has penetrated into the mucous layer is caught in a matching depression ("receptor"), there is an effect on the nerve cell belonging to this cilium. The nerve cell now

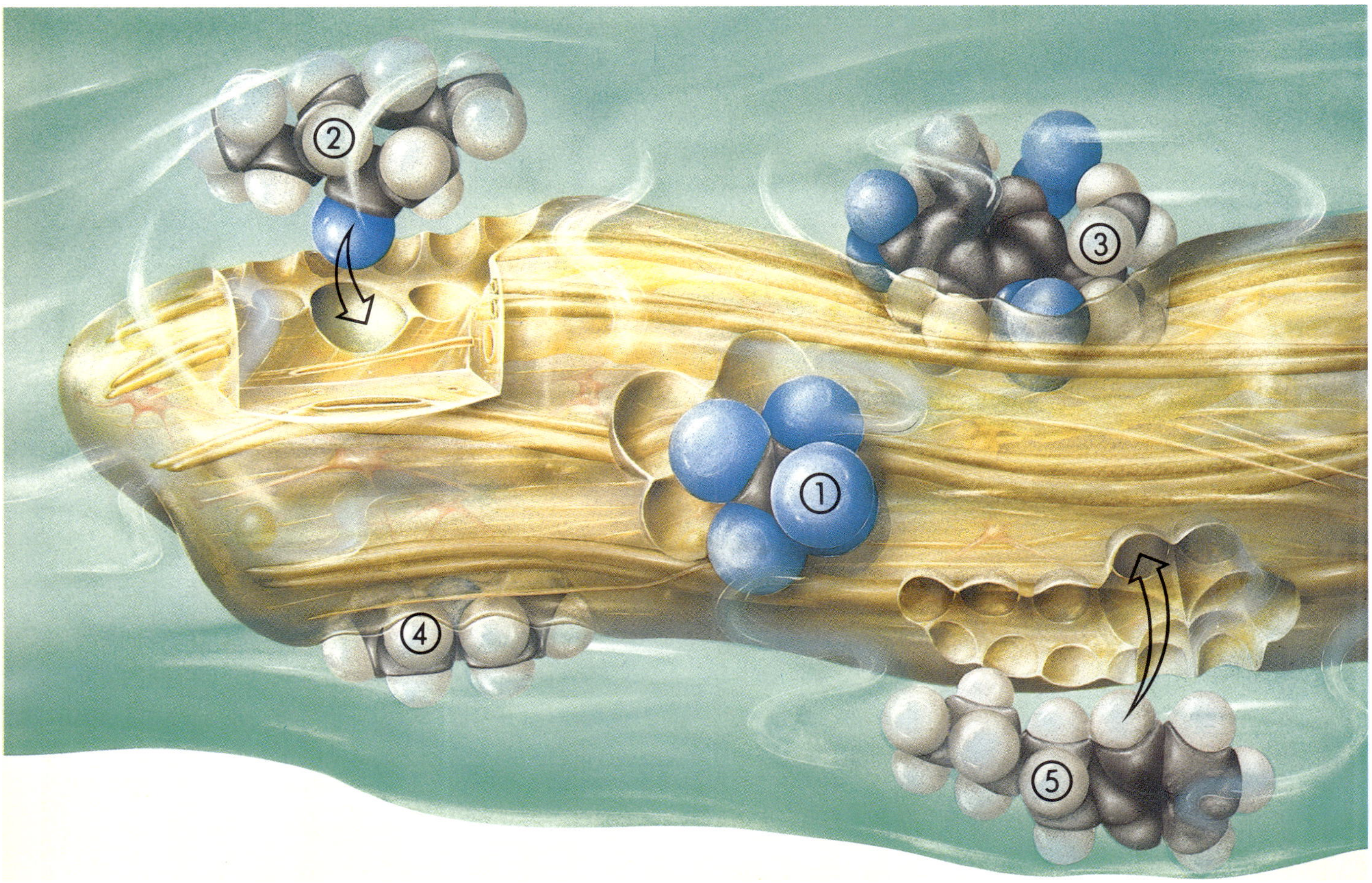

Legend 4
Structures belonging to the olfactory brain, without the surrounding portions of brain, as seen from the top. (1) Spoon-shaped olfactory bulb, (2) Olfactory stalk, (3) Olfactory trigone, (4) Amygdala, (5) Central thalamic ganglion.

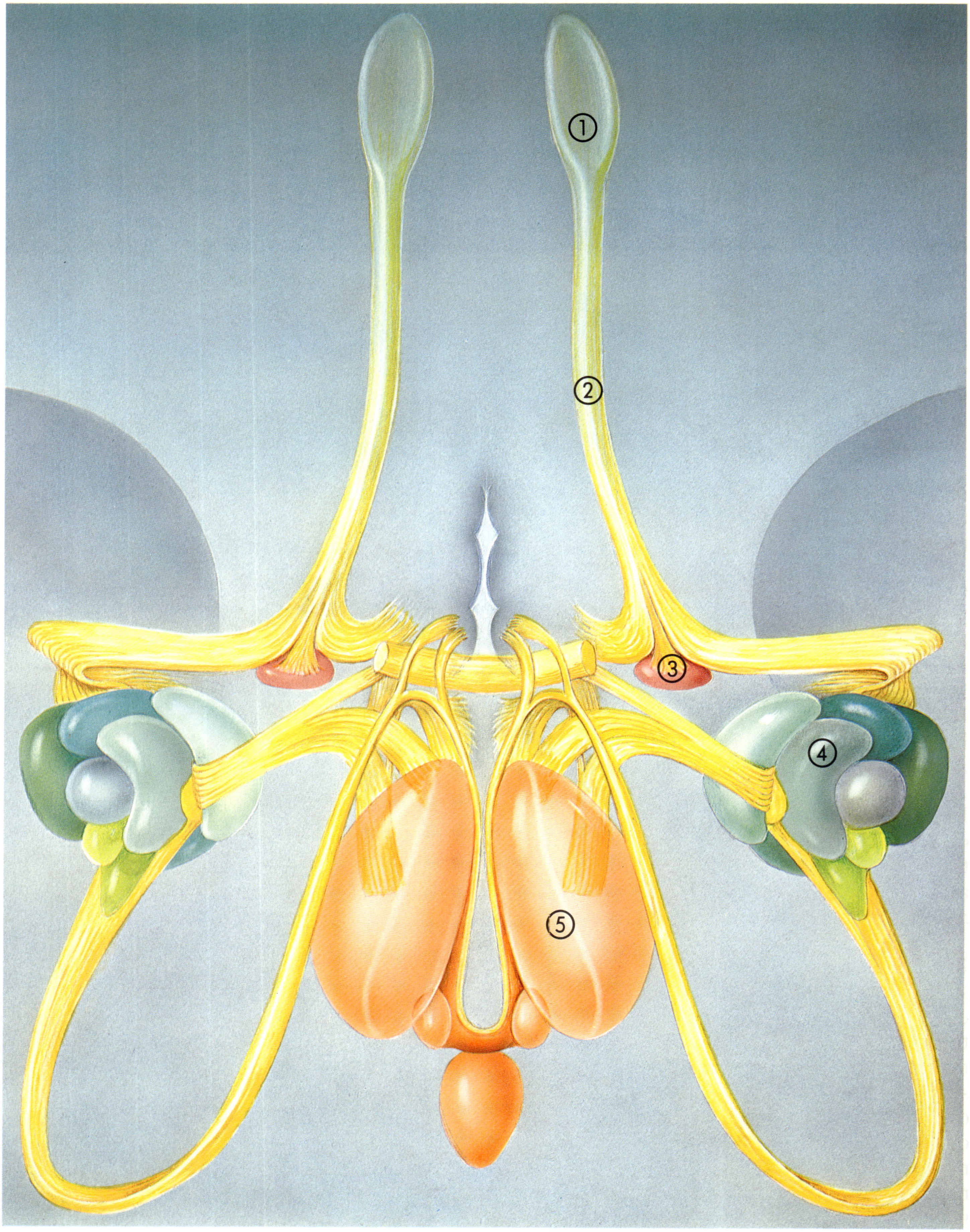

the nerve stimuli are advanced to the extensive ganglia region of the thalamus (Fig. 4) and to other parts of the above-mentioned limbic system, and finally also to the reticular region–the formatio reticularis–of the central brain, which serves as our brain's clock or master switch. The bundles of nerve cell processes that interlink the above brain structures usually do not possess a uniform direction of transmission, but have nerves that travel in the opposite direction. Thus, there is a sort of feedback that permits the input of an odor to be variously modulated and modified via the olfactory mucous membrane.

More or less in the same manner in which we can perceive different colors through our sense of vision, there are very many different odor qualities among smells and fragrances. However, unlike colors, they cannot be classified into a spectrum on the basis of their wavelengths, but are largely determined by the different shape of the odor molecules, on the one hand; and by the shape of the receptors on the olfactory cilia, on the other. According to Amoore's stereochemical theory of olfactory perception, odorants can be differentiated into various classes, each characterized by an example.

Odor class	Characterizing example
Minty	Peppermint
Floral	Roses
Ethereal	Pears
Musky	Musk
Resinous	Camphor
Foul	Rotten eggs
Acrid	Vinegar

Endeavors to provide a distinct classification of odors and fragrances are complicated by the fact that nerve cell processes of a cerebral nerve that also imparts taste perceptions terminate in the nasal mucous membrane, as well.

In the case of human beings, too, olfactory sensitivity is also fairly great. Many substances can even be perceived when diluted billions or even trillions (10^{12}) of times. Studies performed on animals, primarily on moths, where the male of the species can be attracted by the scent of the female over great distances, suggest that contact between only one olfactory cilium and one single odor molecule is sufficient to stimulate an olfactory nerve cell. However stimulation of an olfactory nerve cell is not the same as perception of an odor.

In the case of the moth, there is no noticeable change in the insect's behavior until several hundred olfactory nerve cells have sent identical messages to the next higher neuron regions. Strangely enough, the odor quality that is experienced can also depend upon the concentration of the odorant. An odorant that has a pleasant fragrance in weak concentrations can smell unpleasant in high concentrations. Thus, an excessively strong jasmine fragrance is not very pleasant, and a leaky can of the finest rose oil in the cargo hold of a small airplane can become an annoyance to the entire crew.

If we broadly classify our human existence into the world of the mind and intellect, in which the activity of our cerebral cortex primarily functions, and the world of feelings and emotions, which are controlled by those regions of our brain that developed earlier in the history of the human race, the sense of smell can be said to belong more to the world of feelings and emotions than to the world of mind and intellect. As we experience our environment, the sense of smell can have an exciting, or inhibiting and intimidating effect, can awaken feelings of desire just as easily as aversion, sympathy or rejection. In itself, it is not a scale with absolute, permanently valid graduation; an expectant mother can abhor odors that she normally likes, such as her husband's highly fragrant favorite tobacco. We are not always aware of our olfactory perceptions; we can virtually never describe a specific odor if our "nose tells us" that we don't like somebody. On the other hand, even if olfactory impressions are still present, we can lose our ability to perceive them through acclimatization. A successful perfumer must not only have a well trained nose, he must also possess a high degree of sensitivity for the world of feelings, for moods and wishes, and for the secret desires of human beings.

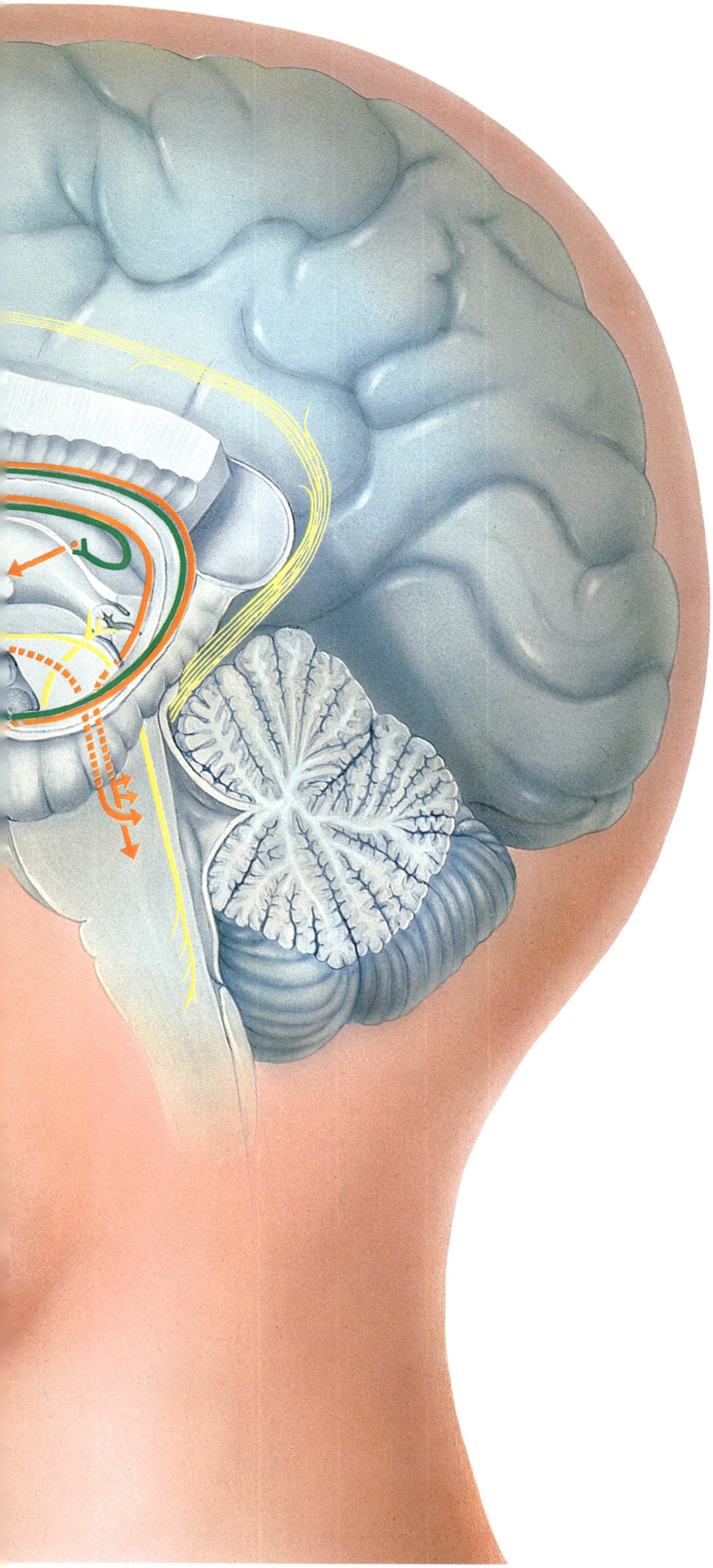

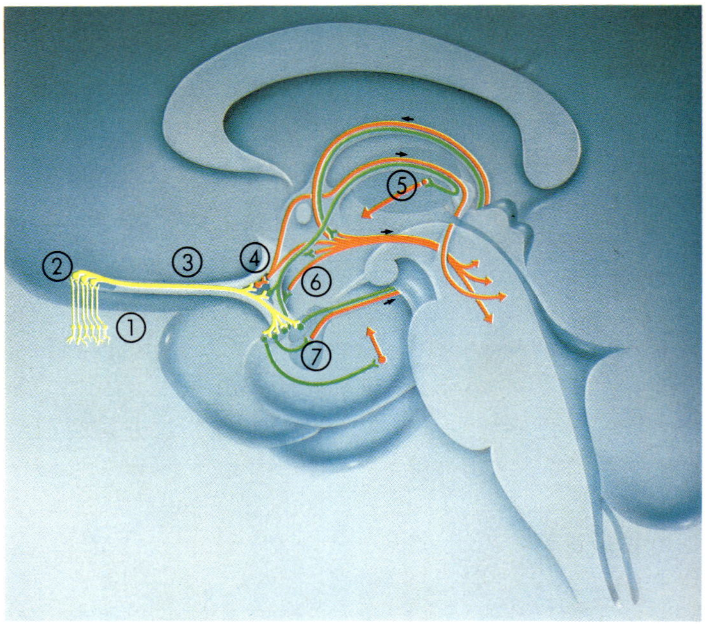

Legend 5
The olfactory system, with its primary and secondary paths to other regions of the brain. (1) Olfactory nerve cells in the nose, (2) Spoon-shaped olfactory bulb, (3) Olfactory tract, (4) Olfactory trigone, (5) Stalk leading to the thalamus, (6) Hypothalamic region, (7) Amygdala.

olfactory mucous membrane enter into contact with the olfactory brain cells located there, with many hundreds terminating on each olfactory brain cell.

It is in the olfactory bulb that the olfactory impressions are initially processed. From there, nerve cell processes are sent in the olfactory stalk to regions of the brain that number among the oldest developed zones of the cerebrum and once rare termed the olfactory brain (Figs. 4 and 6); we know today that the function of these regions is not limited only to the sense of smell, but is also responsible for our feelings. Numerous paths (Fig. 6) link these second-order olfactory centres with the other parts of the sensory system, which today is usually termed the limbic system, at the transition between frontal and temporal lobes.

Most of the bundles of nerve fibres that make up the olfactory stalk, extending from the olfactory bulb to the brain, reach the small, olfactory trigone at the base of the olfactory stalk, nearby cortex regions on the anterior inner surface of the frontal lobe and–some via this, some directly–the nerve cell masses of the amygdala. The above-mentioned structures are the second-order neurons for the nerve stimuli that produce our sense of smell, which are reached from the first-order neurons, the olfactory bulb. From there, sends a signal to the brain. The ciliate structures, with their slender tips (Fig. 7), are designed similarly to other cilia, e.g. those in the respiratory path. The cilia are probably capable of very slow motion and can thus catch matching odor molecules in the mucous film.

In the approx. 8 square centimeters (1 square inch) that make up the combined area of the two olfactory mucous membrane regions, we human beings carry some 10 million olfactory nerve cells (Figs. 2 and 3). In view of the much more highly developed sense of smell of dogs, it is not surprising that their olfactory mucous membrane regions have about twenty times more nerve cells and that this region is some 15 times greater in size in the case of large dogs. The individual fine processes of the nerve cells are bundled in the nasal area and extend through the cribriform plate of the ethmoid bone behind the septum and into the interior of the skull. There, they enter a portion of the brain that can be recognized as the olfactory bulb in human beings (Figs. 4 and 5). In the olfactory bulb, the processes of the nerve cells in the

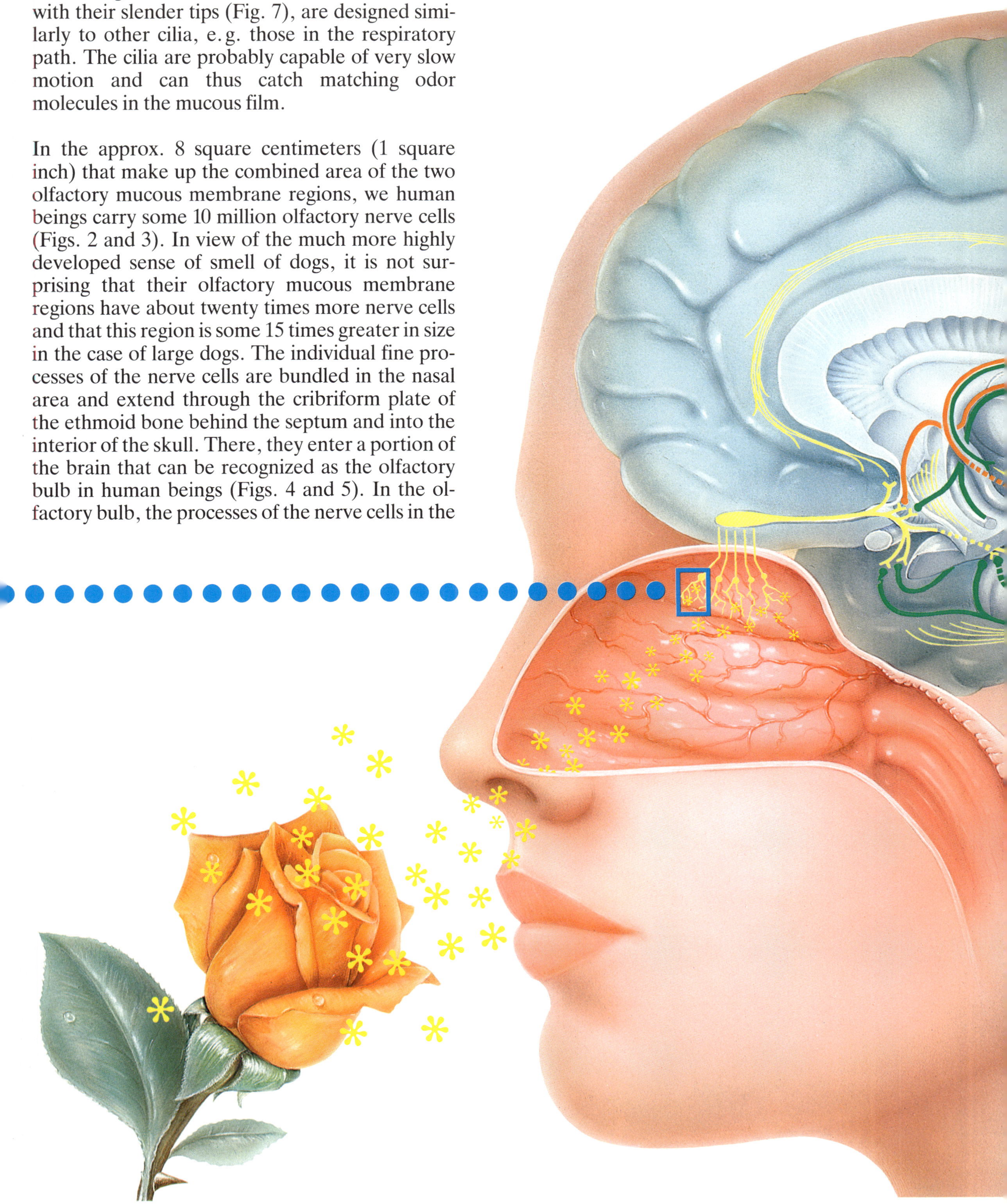

Legend 7
Electron microscope photograph of the olfactory epithelium. The thick, bulb-like protrusions are processes of the olfactory nerve cells. At the bottom can be seen the olfactory cilia, somewhat damaged by being prepared for the slide. Magnification: 2,000 power.

Legend 6
This illustration schematically shows how the olfactory system is linked with the limbic system and the hypothalamus region. Located here are the nervous structures in which feelings are processed and controlled.

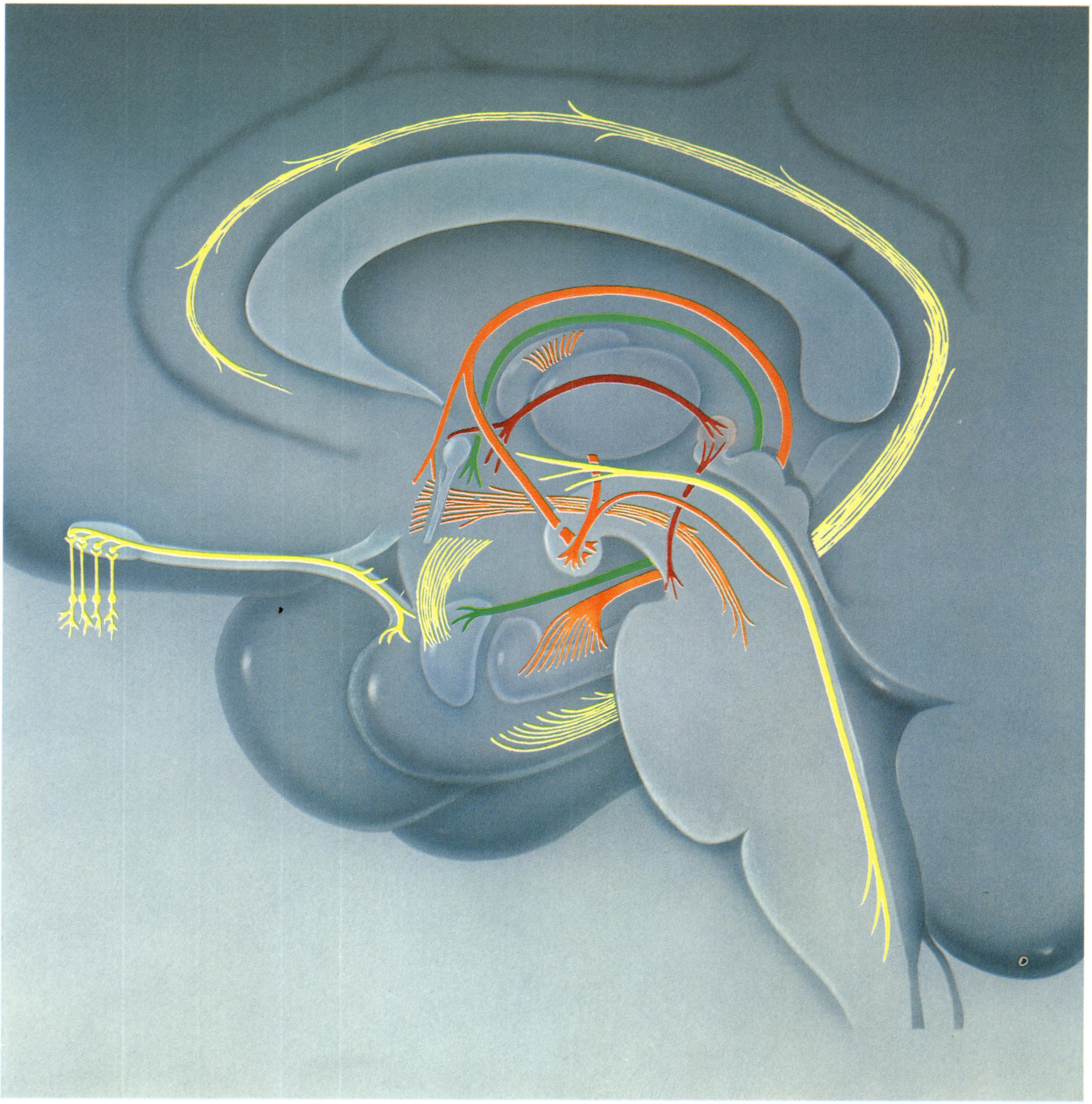

The Perfumer: Profession or Vocation?

A man with normal sexual drives does not remain unaffected for long in a perfume laboratory. The fragrances he experiences there are so erotic, that he stands the chance of being driven to distraction. The foregoing opinion comes from one who should know. He is French Perfumer Pierre Blaizot. And M. Blaizot pleads for understanding: "Let us pity the poor perfumers, who are forced by their profession to live in an atmosphere so dangerous for their health and virtue. Indeed, the Guild does not have the best reputation. But we should not be too critical of these people for their lapses, which are simply an occupational hazard." (From Parfums et Parfumers, Editions du Layet.)

The perfumers themselves, on the other hand, do not describe themselves as especially sensual beings but as members of a very special guild requiring certain knowledge, skills, craftsmanship and routine, as well as a special sort of intuition and creativity. Anyone can call himself a perfumer–nobody holds a patent on the title. But the genuine perfumer is able to compose–from hundreds of fragrance substances–beautiful perfume creations that will sell; and not everyone has the ability to do that.

In France, the perfumer is called Le Nez, "The Nose," which might call to mind some sort of a nose animal, an oversized olfactory organ on top of two feet, sniffing its way around the world. This impression is supported by the standard cliché photograph of the perfumer at work in which he (very seldom is it "she") is shown wholly engrossed in sniffing at a smelling-slip.

That a perfumer must possess a good sense of smell goes without saying. But all healthy people have a good sense of smell. The budding perfumer is not discovered in early childhood playing with scent bottles instead of toy trains.

A perfumer may formerly have been a chemist, laboratory technician, pharmacist, or none of these. He may have a high school education, or he may have gone to a technical school. Or perhaps it was just elementary school. A good education and special training are desirable for this occupation, but not essential. Much more important are artistic talent and a certain sensitivity. It is not necessary that the future perfumer be able to compose, write or paint; but his senses must be open and receptive to beauty, harmony, sounds, pictures, nature and, above all, people. The eroticism referred to in the beginning was not entirely in jest. Only the capacity for great sensual enjoyment, and attunement to feelings such as sexual attraction, can give the perfumer the ability to create something as enticing as a fragrance of quality must be.

The person involved in this profession first came in contact with it in his own environment: in his family life, or as an employee of a perfumery or cosmetic company. Once a person has entered the creative perfumery industry, his elementary education usually lasts about three years. It is an intensive course, in which the trainee learns to know and be able to differentiate between the raw and basic materials, the natural and the synthetic fragrance substances, and their chemical dispositions and fragrance characteristics. He learns the analysis procedures and compounding techniques, not only for the so-called fine perfumery, but also for the broad field of everyday perfuming, from cosmetics to hair and body-care products to household cleansers, and all the rest of the thousands of products which contain fragrances. The total course takes five or six years.

Naturally, the education of the sense of smell and the development of an odor-memory occupy much time. Even the experienced perfumer practices daily. To do this drill, two colleagues usually prepare several smelling-slips. One must then identify the scents on the strips the other has prepared, and vice versa. While the amateur is able to identify a few dozen of the best-known odors, the perfumer has most of the two thousand or so fragrance components he works with in his memory–from the uniform chemical substances via the essential oils to the complete bases and compounds. As an orientation–help, many perfumers build picture-bridges, associating pictures with fragrances–for example: orris = powder, patchouli = oriental market, geranium = Egypt, jasmine = erotic. The odor–memory, the ability to imagine complete fragrance complexes, the ability to recall them and re-arrange them in the mind–this is the true talent of the perfumer. His nose is simply the most important instrument for the job.

Sport or Sin, Carriage or Volcano of Love.

It is not by chance that the name of a perfume usually expresses its personality, mainly by the images it conjures up–images which often have nothing to do with olfaction. In the case of "Quelques Fleurs," those who appreciate flowers will probably imagine a garden or bouquet. But what of "Soir de Paris"–just how does an evening in Paris smell? "White Shoulders?" "Turbulences?" "Risqué?" "Laughter?" or "Grain de Sable?" How do "Risk," "laughter," or a "grain of sand" smell? Of course they have no smell, but they can awaken associations that are translatable into fragrance pictures in our imaginations.

But a fragrance picture is never simple, clear-cut and unequivocal. Just as the rose in the garden does not simply smell of itself. Depending on the time of day and year, there is mixed with the rose perfume the fragrance of dewy grass, or earth, or hay, or the herb bed, and all the other plants in the vicinity.

The visions that a perfume such as "Bois de Boulogne" can conjure up, for example, are practically impossible to describe.

The ingredients to be used in a perfume depend upon the fragrance type it is to be. Should it smell spring-fresh/playful, or oriental/mysterious? The perfumer can pre-formulate such impressions in the form of a recipe–it reads much like a recipe in a cookbook, in fact.

The real precision work is pure craftsmanship –measuring, mixing, trying, rejecting, putting aside, thinking it over. And again and again, smelling. One hundred, two hundred and more fragrance components can go into such a composition.

These components are not only in the perfumer's mind, but also physically in front of him. "Fragrance Organ" is the name given to the array of fragrant materials the perfumer has at his command. Around the turn of the Century, it used to be that the fragrant substances were in fact lined up around the perfumer in tiers, like the keys and registers of a big church organ. Right within the reach of his hands he had the flower oils and absolutes, the herb essences, resins, balsams and drug extracts in bottles, jars and cans.
Today, the fragrance organ looks rather like a modern pharmacy, with more than a thousand little neatly labelled bottles on endless rows of shelves, and refrigerated compartments for the most delicate substances. Inscriptions do not only say "rose," "lily-of-the-valley," and "musk," but also phenyl ethyl dimethyl carbinol, dimethyl benzyl carbinyl butyrate, or cyclopentadecanolide.

The perfumer does not work in this "pharmacy", but in an enclosed, air-conditioned studio. The fragrant substances he selects in this sanctum, are mixed in the laboratory according to his directions, i.e. his formula.

A Fine Starting-Point for a Good Perfume.

The perfumer Horst Maria Faber writes of the complicated work involved in building a perfume in his little book "Tender Breviary of Fragrances." He chooses a chypre fragrance for his example.

"My project begins with the oils of the bergamot, orange and a little mandarin. Or instead, you might use a ready-made eau-de-Cologne oil. The perfumers call this 'top note' or 'Odeur de tête.' Then comes a mixed flower note, mainly represented by a series of jasmine odors. You take a little rose and a few synthetic rose substances, and–please pay attention: the fresh and intense tonality of the citrus oils would clash with the ensuing floral notes if I did not create a 'bridge' with lavender and a little coriander oil and Brazilian rosewood oil. A similar bridge again becomes necessary to lead the flower notes to the actual 'fond' of the perfume, which is its base. There, you meet quite a few old friends: East Indian sandalwood oil, vetiver oil, and the somewhat musty patchouli oil, which takes on a pleasant smell only when it is very 'vieux.'
In addition come several orris ingredients partially of natural origin, partially synthetic, such as ionone and methyl-ionone. All these together in the proper proportions give an excellent platform for a good Chypre. Now add the back-notes, which will cling to clothes and laundry long after the 'overture' of cologne and florals has been forgotten: Oak Moss Extract and synthetic Musk in quantity, and above all some Coumarin, Heliotropin, Vanillin, resins and balsams. . . ."

So much for the sketch of a classic perfume, one which has given its name to a whole fragrance family, the chypre notes. Nearly every woman has encountered Chypre at one time or another, whether under the guise of "Ma Griffe" or "Femme", "Miss Dior" or "Cabochard".

The Fireworks Display in the Bottle.

As we've seen, fragrance is not a fixed, immutable object, but a happening that develops over time.

Applied to the skin, the perfume first "explodes" like a firework, then calms itself, comes to its senses and then stays on for hours, revealing its true self.

What your sense of smell first experiences in a perfume is known as the "top-note" or the "initial impact." Then, after a few minutes, comes the "bouquet," the "heart," or "middle note," which gradually blends into the "back-note," also known as the "base" or "fond" in French.

The evaporation-process of a perfume could be visualised as a rocket with 3 stages being ignited in succession. The perfume components are placed in order from the top to the base, according to their volatility–from the very light ones to the heavy, least volatile; from the tingling fresh to the warmer ones. The top-note could be fresh citrus and light, volatile green notes, aldehyde and fruity esters. For the bouquet of the middle note, there could be accords of several flower notes. And for the base, woody notes, patchouly, moss, balsamic and animal notes, coumarin and vanillin.

What is often called the "fixative" could also be called the "fragrance retarder." These retarders are, in contrast to many essential oils, slow to evaporate, because of low vapor pressure. The following fragrance components are in this category: Benzoin Siam Resin, Olibanum Resin, Ethyl Vanillin, Heliotropine, beta-Methyl Naphthyl Ketone. They range in consistency from highly viscous to crystalline. Their function is to bind together the ingredients of the composition and lead them from the top to the base. So, unlike that three-stage rocket, a perfume does not break up into three stages, each of which smells differently and leaves the others behind–but the head contains some of the heart, and the base retains some of the middle-note fulness. If a perfume is well-balanced, changes during the evaporation-process are as imperceptible as the movement of the hands of a clock. A perfume that falls apart, in stages, is badly formulated.

A perfume should be harmonious, rounded-off, balanced–but not boring. It should be reminiscent of natural and familiar odors, but also bring new and surprising ones. It should be a mirror of the contemporary aesthetic feeling, but not simply reflect some passing fashion. It should please. But how do you please millions of women?

A Nose for Millions.

A perfume that is a Big International Success is as rare as a faultless diamond. It might be created in the space of six months, but this is exceptional. Usually the perfumer works on a project for years before he is satisfied. Though his success may end up on the dressing tables of millions of women, he normally remains anonymous. There are exceptions, for example Jean-Paul Guerlain, who composed fragrances such as "Chamade" and "Nahema" for his famous family company. Once in a while, there is also a loner, such as the great Edmond Roudnitska, who created "Femme" for Rochas, and most of the famous perfumes of the house Dior. But normally a perfumer works as part of a team in a big company, and his name remains unknown to the outside world. You might liken the perfumer's nose to the tip of the ice-berg, the submerged bulk being the prosaic aspects of the fragrance industry–raw material procurement, research, analysis and applications.

When the perfumer meets a woman who wears his creation, he knows it does not belong to him anymore. If it is successful, it is a foregone conclusion that every woman will claim it for her own. For that is exactly what is success for a perfume.

Fragrance Families: a Classification of Perfumes

No perfume is unique. No matter how original or exclusive a creation may seem to be, its fragrance is always compared to other related ones, to describe both similarities and differences. It has become necessary in the perfume industry to put the existing extrait-perfumes in a clear order. First, to facilitate better communication between the specialists who work with fragrances; second, as a help to interested amateurs–perfume users who may have difficulties in describing their fragrance preferences and requirements.

It is no coincidence that perfumes are grouped into families. The living and ever-growing order of kinships, clans and dynasties closely corresponds to the relationships which exist among the perfumes of the world.

In every perfume family there are beloved old standards and classics, especially prominent members, which have also influenced other families. Therefore, any description of fragrance families must include the family-tree, the genealogy of the extrait-perfumes.

The names of the families are: *Green, Floral, Aldehydic, Chypre, Oriental, Tobacco-Leather, Fougère.*

The Family of the Green Notes.

Green Notes, the youngest branch of the family tree, have their roots in a solid familiar background. On the one hand they are related to the old established floral notes; on the other, to the clan of the chypre fragrances. The similarities are so strong that it is not always possible to differentiate the offspring from the relatives. All in all, "green" describes the fragrance conception of leaves, meadows, grass. The family is subdivided into *fresh* and *balsamic,* in light and cool notes of lemon and herbs of the green direction; and also in the soft, warm notes of natural odorants–used mainly in perfumes of the chypre type.

The modern, fresh line is inspired by floral notes. Here is an abundance of interesting successors to the original Vent Vert/Balmain (1945), Weil de Weil/Weil (1971), Espiègle/Atkinsons (1973), and Silences/Jacomo (1979); and also, Chanel N° 19/Chanel (1971) and its successors.

Among the balsamic green notes is Alliage/Lauder (1972) the most significant perfume, the astringently fresh floralcy of which–in connection with a mossy, forest-like background–has influenced many perfumes which have followed it.

The Family of the Floral Notes.

This is a large family, both in tradition and in the number of its members. The floral notes constitute the largest odorant-reservoir in the perfume industry. They are sub-divided into: *fruity-fresh, floral-fresh, floral,* and *floral-sweet.* Among the fruity-fresh notes, there are new-style perfumes, such as Amazone/Hermès (1975), with an original touch of cassis in the top-note; and Lauren/Warner-Lauren (1978), with its pineapple-like freshness. Successors are Métal/Rabanne (1979), and Anaïs Anaïs/Cacharel (1979).

The floral-fresh branch is led by Muguet des Bois/Coty (1936). In its succession are the classics from Dior, such as Diorissimo (1956) and Diorella (1972); and also, Fidji/Laroche (1966). Fidji initiated its own fragrance trend, which, in turn, influenced Charlie/Revlon (1973).

The purely floral fragrances, with world-famous compositions such as Quelques Fleurs/Houbigant (1912), Joy/Patou (1935), and L'Air du Temps/Ricci (1948), staged an impressive comeback in the Seventies.

They inspired the perfumers to create new kinds of floral accords such as Tatiana/v. Fürstenberg (1975), or Fantasque/Féraud-Avon (1981).

The perfumes of the floral-sweet direction contain comparatively large amounts of fixatives for a long-lasting effect, and for this reason their direction is almost Oriental or exotic-floral. L'Origan/Coty (1905), started this line which led via Oscar de la Renta/Stern (1976) to remarkable new creations in the Eighties. The tuberose-accented creations such as Chloé/Lagerfeld (1975) were also of great importance.

The Family of the Aldehydic Notes.

The aldehydic notes have an especially strong diffusive quality, which makes them difficult to compare with fragrances of Nature. This family, still predominantly fostered by French perfumers, has a *woody-powdery* direction in addition to the basic aldehydic-floral characteristics. The perfume that gave its name to the whole family is Chanel N°5/Chanel (1921), the first perfume of importance to confess that it contained purely synthetic fragrance materials–the aldehydes. Its influence on compositions ranging from Arpège/Lanvin (1927), Madame Rochas/Rochas (1960), Climat/Lancôme (1968), to White Linen/Lauder (1978), to the latest creations of the Eighties, is unmistakable. The sub-division *aldehydic-floral-woody-powdery* notes reveals its characteristics at first glance–through its name.

From Tosca/4711 (1921), through Bois des Îles/Chanel (1926), Calèche/Hermès (1961), Calandre/Rabanne (1968), to Chamade/Guerlain (1970), and further creations, they are all highly original and complex, sometimes heavy fragrances.

The Family of the Chypre Notes.

The Chypre accord is based on the harmonic contrast of bergamot-accented freshness with an oak moss foundation. The family name originates from Chypre/Coty (1917), the basic concept of which has been changed and expanded in many different ways. Today, there are equal numbers of chypre notes in both women's and men's fragrances, they hold a special position for that reason. Chypre is sub-divided into *fresh-mossy-aldehydic, floral-mossy-animalic,* and *mossy-fruity* sectors.

The fresh-mossy-aldehydic chypre type is given depth by dry, algae, and woody-mossy accents. In the succession of the original Chypre, which is no longer produced, we find creations such as Mitsouko/Guerlain (1919), Crêpe de Chine/Milot (1925), and Ma Griffe/Carven (1944). During the Seventies, there were no new developments of any importance, but the branch has become active again in the Eighties, partly because of renewed interest in Oriental notes.

The floral-mossy-animalic chypre notes are very numerous and varied. Floral, spicy, leathery and animalic modifiers can all be added to the basic accord of Citrus and Oak Moss. Miss Dior/Dior (1947), Intimate/Revlon (1955), and above all Cabochard/Grès (1958), are the most successful representatives of this line.

The rather heavy, earthy ingredients of the mossy-fruity chypre perfumes are rounded off by sweet-fruity nuances. The protagonist and–up to now–most prominent example of this direction is Femme/Rochas (1942). For a long time, this perfume had no successor, but the theme has become popular again in the Eighties.

The Family of the Oriental Notes.

This is the family with the longest unbroken tradition. Peculiar to it is a distinctive heavy sweetness, achieved by the use of various resins and animalic fixatives. Jicky/Guerlain (1889), is the oldest perfume still on the market. Its successors, including Shalimar/Guerlain (1925), and Vol de Nuit/Guerlain (1945) influenced new developments of the Eighties. With Shocking/Schiaparelli (1935), Youth Dew/Lauder (1952), and finally, Opium/St. Laurent (1977), "The Orientals" began a whole new, modern career.

The Family of the Tobacco-Leather Notes.

The fantasy accords with tarry and animal elements which differentiate this family from the others belong to the area of the masculine fragrances. Tabac Blond/Caron (1919), and Cuir de Russie/Chanel (1924), were for many years the only representatives of importance. But a revival of this trend is to be expected.

The Family of the Fougère Notes.

Like tobacco-leather, this is primarily a masculine fragrance type, with herbaceous-fresh notes shaped by lavender and moss components, and coumarin in the background. The family takes its name from Fougère Royale/Houbigant (1882). Originally conceived for women's perfumes, this odor-type has moved almost completely into the masculine domain. There is also a revival to be expected here in the area of lighter alcoholic perfumery such as eau de toilette.

The Fragrance Families

A Classification of Fragrances According to Their Dominant Characteristics.

Each of the perfumes illustrated here is the oldest in its category and can be considered as the proto-type of its line.

al
ery
fresh-mossy
aldehydic
Chypre Notes
floral-mossy
animalic
mossy-fruity
Oriental
Notes
Tobacco
Notes
Leather
Notes
Fougère
Notes
Cuir de Russie
Chanel 1924
CUIR DE RUSSIE
CHANEL
PARIS
Bandit
Piquet 1944
bandit
DE
ROBERT
PIGUET
TABAC
BLOND
CARON
Tabac Blond
Caron 1919
Femme
Rochas 1942
Jicky
Guerlain 1889
Fougère Royale
Houbigant 1882
osca
711 1921
Chypre
Coty 1917

Perfumery Lore at a Glance: from "Absolutes" to "Woody"

Absolutes
Absolutes are natural fragrance materials which are extracted from various plant parts. First, the concrete is extracted by means of a solvent. Then, the undissolved waxes are removed with alcohol. Absolutes are products of high quality, and very expensive because of the low yield.

Accord
Accords are combinations of various single odors which blend to produce new fragrance effects. The number of ingredients in an accord may run from 2 to several hundred. Both simple and complex accords may be used as components for perfume compositions.

Aerosol
Aerosol is the spraying and/or foaming of liquid or solid materials by propellant agents from pressurized cans.

Agrumen Oils
Agrumen oils is the collective term for the essential oils of bergamot, lemon, grape-fruit, lime, mandarin, orange and bitter orange, generally known as citrus oils.

Alcohol
Alcohol is used in the perfume industry as a solvent for the production of lotions. An often-used alcohol is ethyl alcohol.

Aldehydic
Aldehydic is the term for the odor-effect produced by the use of short-chain aliphatic aldehydes. This effect can be described as fatty, watery, tallowy, or even "snuffed candle". When concentrated, aldehydes are extremely powerful and pungent. Aldehydes are used in all perfume types, especially those which feature elegant feminine notes.

Animalic
As the name indicates, animalic odorants originate in the animal kingdom. They are extracts from animal secretions or synthetic imitations of these. They can also be extracts of plants which have similar fragrance characteristics, or even single chemicals.
The best-known products of the animal kingdom in the perfume industry are civet, musk, castoreum and amber. In concentrated form, they often smell unpleasant and obtrusive. Correctly diluted, however, they are an irreplaceable ingredient in many perfumes, giving them warmth and fullness.

Anosmia
Anosmia is the inability to smell. Some people have a selective or partial anosmia, others a complete anosmia.

Aromatic
Aromatic has become an obsolete term in perfumery. In former times, it was used to describe the fragrance impression of sweet balsam.

Association
Association is the connecting of sensory impressions, moods or incidents with odors. A scent has the ability to bring a situation you have experienced from the deepest recesses of your mind, to the absolute forefront.

Back Notes
The back note is the third and last phase of a perfume's life on the skin, or evaporation. It contains the lasting ingredients, such as woods, resins, animal and crystalline substances. In heavy perfumes (chypre and Oriental notes, for instance) the back note is so strongly accented that it is discernible in the top-note, or first impression.

Balsamic
Balsamic is a fragrance impression that can be described as sweet, soft and warm. Basically, balsamic notes result from the use of balsams and resins in perfume compositions. The Oriental perfumes in particular are characterised by balsamic ingredients.

Balsams
Balsams are viscous secretions of plants that emerge when the plant's outer layers are injured. Unlike the resinoids, balsams do not require an extraction process prior to their being used in perfumery.

Bitter
Bitter is the fragrance impression that corresponds to bitterness in terms of taste. It is produced by a combination of roots (such as vetiver), herbs (such as wormwood), animal notes (such as in leather) and others. Bitter accents are to be found mainly in masculine fragrances.

Body
See "Heart".

Bouquet
Bouquet is a mixture of various floral notes. Often, the bouquet is the most important ingredient of the middle-note in a perfume. Bouquetting is the embellishment, harmonizing and rounding-off of a composition.

Camphoraceous
Camphoraceous notes have a fresh, clean, medicinal smell. In Nature, they are present in lavandin, rosemary, and conifer oils, among others. These notes are widely used in the perfume industry in the perfuming of bath products.

Chypre
Today, chypre is the collective term for a group of perfumes which get their character through the combination of a fresh Eau de Cologne-like top-note and a foundation that comprises as main elements oak moss, labdanum and patchouly. Many warm, erotic, sensual perfumes belong to the chypre family. One of the classics is "Chypre/Coty," a perfume that has been on the market since the beginning of this century.

Citrus Notes
Citrus notes have a fresh, light character. They come from the family of the agrumen oils (bergamot, lemon, lime, mandarin, orange, bitter orange). In addition to these there are a number of synthetic substances that have the fresh character of citrus in different variations.

Coniferous
Coniferous is the odor impression that is reminiscent of pine, spruce, juniper and others of this family. Conifer oils are mainly used in bath products and in masculine fragrances.

Composition (or Compound)
Every perfume is a composition of harmoniously adjusted individual components. The characteristics of the individual ingredients are used to create a new and unique overall picture, in combination, in which the individual characteristics of the ingredients recede in favor of the combination, but at the same time, are not themselves obliterated.

Creation
Creation is the making of a new perfume oil composition. Creative work is an art which is nevertheless influenced by economic considerations, so perhaps "skilled craft" would be a more appropriate description.

Crystalline
A certain amount of crystalline fragrance substances can be dissolved, without problems, in the liquid ingredients of a perfume oil. In general, they possess good fixative properties. But too high a percentage of crystalline substances, and especially low temperatures, can cause what is known as "crystallizing-out"–the appearance of crystals in the liquid.

Distillation
Distillation by steam is the most commonly-used process for the production of essential oils. In this procedure, steam flows through the distillation material and sweeps the essential oils along with it. After cooling, the distillation water is separated from the essential oil in so-called Florentine flasks.

Dry
Dry is the term for the odor-effect, the reverse of "sweet" or "warm", achieved through the use of ingredients such as woods, mosses, herbs and so on. Dry notes are used mainly in masculine perfumes and are particularly useful as fresh, daytime fragrances.

Eau de Cologne
Eau de Cologne is a solution of approximately 3% to 5% perfume oil in an alcohol/water mix.

The classical "eau de Cologne" is a composition of fresh, light, volatile essential oils (predominantly citrus oils) which contains few, if any, fixatives. Eau de Cologne is intended to be refreshing, and has a limited perfuming effect.

Eau de Parfum
Eau de Parfum is a solution of perfume oil, (15%–18%) in alcohol (85% to 82%).

Eau de Toilette
Eau de Toilette is a solution of 4%–8% perfume oil in alcohol.

Earthy
Earthy is the adjective used to describe the fragrance impression of earth, forest-soil, mold, dust, etc. Vetiver and patchouly are well-known essential oils possessing earthy characteristics. Earthy accents in perfumes are not pronounced, always subtle.

Encapsulation
To encapsulate, in the perfume industry sense of the word, means to enclose perfume oils in tiny gelatine capsules. These capsules can be applied to the skin together with an alcoholic perfume. When the skin is rubbed, the capsules are broken and the scent of the oil is released, "renewing" the perfume. Tests involving this method of perfuming have been made with textiles.

Enfleurage
Enfleurage is a process for the extraction of valuable plant extracts. Plates of glass, covered on both sides with animal fat into which blossoms have been pressed, are placed on wooden frames. Spent blossoms are constantly replaced until the fat is saturated with fragrance substance. Then, the blossom oil is separated from the fat through extraction. This procedure is rarely used today, because it is so costly.

Erotic
Perfumes which, in addition to their general pleasing and harmonious qualities, are also accented with warm animal notes in combination with certain flower oils, are said to have an erotic effect. Much depends on the user, though–and the circumstances!

Essence Concrète
Essence concrète–"concrete" in English–, is extracted from different plant parts through the use of solvents. In addition to the essential oil, the "concrete" contains a number of waxes not soluble in alcohol. Essence concrète is mostly used in soap perfume because of its good substantivity and the insoluble substances. In alcoholic perfumery, the so-called "absolutes" are used. These are prepared by eliminating the waxes from "concretes", thereby improving solubility and strength of odor.

Essences
Essences are alcoholic or aqueous plant extracts. They are hardly ever used in the perfume industry today, but they are widely employed in the cosmetic and flavor industries.

Essential Oils
Essential (or Ethereal) oils are extracted from various plant parts through pressing or steam distillation. They are natural mixtures of various chemical substances. Unlike fatty oils, they evaporate without leaving a residue.

Evaluation
Evaluation means the selection of fragrances, for a specific purpose, from a number of available alternatives. In recent years a whole new profession has emerged in the perfume industry consisting of experts who specialize in this work.

They are known as evaluators. Evaluation is work which requires not only talent, knowledge and training, but also a feeling for what the market wants at any particular time. The evaluators (women, in many instances) are an important link between the creators and the users of perfumes.

Expression
Expression is an especially mild process for the extraction of essential oils, used in cases where steam distillation would modify or damage the end-product. Expression is used mainly for the extraction of citrus oils.

Extraction
Extraction is the process of removing raw materials from plant and animal substances through the use of various solvents. Most of the valuable natural fragrance materials are produced in this manner.

Extrait
Extrait, generally, means the most concentrated form of perfume sold over the counter. It is a solution of 15%–30% perfume oil in high-grade alcohol.

Fatty
Fatty is the name for the odor that suggests oil, lard, wax and so on. In small doses, these notes are reminiscent of the smell of human skin. They can therefore contribute to the erotic effect of a perfume.

Feminine
In the context of perfumery, femininity is a quality that is judged subjectively. Any perfume that underlines the womanly attributes of its wearer can be termed feminine, though much depends on the rapport between fragrance and user.

Perfumes with floral accents are generally considered to be especially feminine. Of course, this does not prevent women from using scents that are considered to have a masculine character, but the majority uses those with feminine characteristics.

Fixing or Fixation
Fixing is a process that promotes the retention of the fragrance on the skin as long as possible. To achieve this, heavy, non-volatile substances are used which develop their full fragrance intensities only very slowly, and maintain them for longer periods. Substances are also used for this purpose which have no strong odors of their own, but have the ability to make other fragrances last longer. Good substantivity is a characteristic of every well-constructed fragrance composition.

It should be noted that an excessive amount of fixative in a perfume is no guarantee of good retention, because substances can hinder one another in their fragrance diffusion.

Floral
Today, over half of the branded perfumes are characterized by the adjective "floral." They contain well-defined flower notes (lily of the valley, for example, as in Diorissimo/Dior), or a whole bouquet of floral effects, as in Quelques Fleurs/Houbigant. As a matter of fact, all perfumes contain floral notes in some quantity.

Floral-Fruity
Floral-fruity perfumes are those that have a noticeably fruity element, especially in the top-note, as an accessory to the main theme, which is always floral. Excesses of fruitiness are to be avoided as they would suggest food-flavor rather than perfume.

Formula
The formula of a perfume compound gives full details of its components–quality and quantity-wise–and is used like a recipe for preparing the mixture. To formulate a perfume can take months, years even, and many experiments. Formulas, holding a company's Know How and ipso facto capital, are jealously-guarded trade secrets, sometimes handed down from generation to generation.

Fougère
Fougère is a fantasy term in perfumery. It is the name of a combination of fresh herbaceous lavender notes on a mossy foundation. Fougère notes occur in many fantasy fragrances–especially in masculine perfumes.

Fragrance Blotters, or Smelling Strips
Fragrance blotters, or smelling strips are narrow strips of absorbent paper about 15 centimeters long with which scent samples are taken and smelled. On smelling strips, the evaporation of fragrance materials and perfume oils can be observed in the different phases they go through. Final judgement of a perfume must always be made on the skin, however.

Fragrance Components (or Ingredients)
All the materials which the perfumers put together to form a perfume composition are known as fragrance components. These are uniform chemical substances, natural products and simple or complex mixtures – the so-called bases and specialties.

Fragrance Diffusion
Fragrance development is the general behavioral pattern of a perfume in the hands of its user. A good perfume should perform three functions. These are:
a) Immediate impact on opening the bottle.
b) Noticeable emanation from the skin in all phases of fragrance evaporation.
c) Noticeable scent in the area that surrounds the user.

Fragrance Material Industry
The fragrance material industry is a branch of the chemical industry. This industry includes the producers of natural and synthetic fragrance materials and perfume oils. The fragrance material industry is a supplier to the manufacturers of perfumes, cosmetics and other products of this nature.

Fragrance Materials, Natural
Natural fragrance materials are products of plant

and animal origin, extracted by different processes. Some examples are essential oils, absolutes, concretes, resins, balsams and tinctures.

Fragrance Materials, Synthetic
Synthetic fragrance materials are produced from chemical raw materials. Half-synthetic fragrance materials are chemically-processed isolates from natural products.

Fragrance Organ or Perfumer's Organ
The fragrance organ is the term for the working-area of the perfumer, in which the fragrance materials he uses in his work are arranged around him in tiers, like the pipes of an organ around an organist.

Fresh
"Fresh", in relation to the effect of a perfume, is a subjective feeling that can be caused by different fragrance impressions. In European regions, freshness is generally associated with lemon, lavender and green notes, light floral components: light and clear elements, mostly. In other regions, North America for example, sweet and powdery perfumes are also considered as fresh.

Fruity
Fruity is the fragrance impression of natural fruit odors, such as raspberry, apple, plum, etc. Their usual role in perfumery is only to produce nuances. However, single fruit odors do become fashionable, from time to time, in such products as shampoos. Exaggerated doses of fruit notes give perfumes the effect of being "edible," i.e. the opposite of erotic.

Gas Chromatograph
The gas chromatograph is an instrument for the analysis of organic chemical mixtures. In a spiral glass or metal column, packed with porous material, the various components are separated according to physical properties such as polarity and vapor pressure. The signals received are amplified and, with the help of a printer, printed onto a chromatogram.

Grasse
Grasse is a town in Southern France, in the vicinity of Cannes, behind the Riviera. The town has a Mediterranean climate, which is well-suited for the cultivation of plants that produce perfume raw materials. For many years, Grasse has been one of the principal centers in the world for the production of natural fragrance components.

Green
Green is the general term for the odors of grass, leaves, stems and so on. Green fragrances exist in many different nuances. They are widely used in perfumery for the purpose of giving special accents to top-notes.

Harmony
Harmony is the tuning of all the components of a perfume so that no single element in any phase of the fragrance evaporation becomes so prominent that it could be considered unpleasant.

It is easy to achieve harmony between similar-smelling substances but quite difficult between the contrasting elements which are often introduced into a perfume to give originality and character. Bringing these warring elements into harmony is a challenging task for the perfumer.

Hay-like
Hay-like notes are used mainly in "Nature" fragrances, in different ranges of application–for instance, in medicinal bath-products. Masculine perfumes also can contain hay-like components (Fougère). The synthetic substance with a hay-like odor that is most important in the industry is coumarin.

Herbaceous
Many fragrance substances have herbaceous components, and are reminiscent of herbs and drugs. Well-known, and often-used examples are mugwort, sage, rosemary and lavender.

Herbaceous accents are widely used in masculine perfumes.

"Heart" or Middle Note
The heart is the second, middle phase of a perfume's fragrance evaporation, occuring after the top note fades away. It is mainly produced by floral, spicy or woody components and represents, as its name indicates, the heart of the perfume.

Heavy
Fragrances in which the least-volatile ingredients such as mosses and animal notes dominate are called heavy perfumes. Since these ingredients are part of the top note, a heavy perfume can be identified as such at first impact. Heavy substances are used predominantly in chypre notes and Oriental notes.

Infusion
Infusion is the production of flower oils by extraction at 65 degrees centigrade with the use of alcohol.

Ingredient
An ingredient is one of the parts that go to make up a mixture.

Intensity
The intensity, or fragrance strength of a perfume compound is dependent on the strength of the individual ingredients, and the skill that is exerted in the blending of these components.

Jasminey
Perfumes that contain the scent of the jasmine blossom as a principal component are termed jasminey. There are many interpretations of the jasmine note, which range from natural blossom odors to stylised, fantasy bases.

Lavender Oil
Lavender oil possesses a dry-fresh, herbaceous odor. It is used in many perfumes, especially in masculine notes.

Leather Notes
Leather notes, as well as tobacco notes, play a significant part in the masculine perfumes. Both natural expressions and fantasy interpretations of this theme exist and are used in the perfume industry. Leather notes also play a part in feminine perfumes; for instance, in the chypre family.

Light
Light perfume notes owe their character mainly to fresh, citrus, floral, fruity and green components. They contain practically no sweet, balsamic or sultry elements. It used to be difficult, but today, the perfumer has the means to give good fixation also to perfumes of light character.

Maceration
Maceration is a procedure by which blossom oils are extracted with the aid of warm fats. It is similar to enfleurage.

Masculine
Masculine, as well as feminine, in the perfume sense, are subjective fragrance impressions. Normally, the term “masculine,” is applied to fragrance favored by men. These contain dry notes of leather, tobacco, herbs, spices, mosses and woods. They are generally less floral than feminine perfumes, and often contain a high percentage of fresh ingredients. In the past few years, masculine and feminine notes have come closer and closer together.

Maturity
A perfume must mature for four to eight weeks before it can go on sale. This time is necessary to allow the individual ingredients to blend, bringing the fragrance to its full development.

Metallic
Metallic notes are used in perfumes to produce cool, clean effects. They are used in nuances, and very seldom occur dominantly.

Mint-Like
Fragrance notes which are reminiscent of peppermint and spearmint are used in perfumery to produce special fresh effects in the top-note.

Mixing Plant
The stage in the production of perfume in which the concentrated perfume oils are mixed, on a large scale, according to the perfumer’s recipe, is known as the mixing or compounding plant.

Modifying
Modifying means varying an existing, basic fragrance theme by changing some ingredients or introducing new, additional nuances. The result may impress a lay-person as a new perfume, but the expert will recognize a mere variation on an old theme.

Mossy
Mossy odors of different kinds of tree mosses (especially oak moss) play an important part in nearly all perfume types. They are of special significance in the chypre notes. Mossy nuances are very complex and can have, besides the basic moss element, algae-like, leathery, woody and other characteristics. Their especially good fixing qualities, as well as their ability to give fragrances substance and depth, make them indispensable.

Musk
Musk is a secretion of the musk deer. The material extracted from musk-sacs has a strong animal-smell. The natural product, as well as chemicals with musk-like odors, are of great significance in the perfume industry. They give

perfumes a warm, erotic note and have outstanding fixing characteristics. Perfumes that are based on musk notes are especially subject to fashionable trends.

Napthalene-Like
Napthalene-like notes are reminiscent of mothballs. These odors are found in animal products. Their significance in perfumery is minimal.

Narcotic
"Narcotic" perfumes often contain high percentages of heavy blossom fragrances (jasmine, tuberose, for example), and animalic components.
The narcotic effect of natural flower scents is at its peak at the time when they fade. Careful dosing of "narcotic" components is necessary if a perfume is not to have an obtrusive or tiring effect.

Nuancer
Nuancers are fragrance materials that are not the main fragrance carriers in a composition, but are used to support and round these carriers off, or produce special effects which contribute to the all-around picture of a fragrance.

Obtrusive
Perfumes can be obtrusive when they are used to excess. Similarly, an ingredient can obtrude if the fragrance composition is not well balanced.

Oriental
Oriental is the term for perfumes containing ingredients that are reminiscent of fragrances from the East. Such ingredients can be exotic blossom notes, spices, balsams, resins, and animalic components. The character of the Oriental perfumes is such that they are mostly used as so-called winter or evening perfumes.

Perfume
Perfume – "per fumum" – comes from the Latin, meaning "through the smoke." In ancient times, fragrant resins were burned as incense offerings– that was the origin. Today, we understand perfume to be a solution containing 15% to 30% perfume oils and 85% to 70% alcohol, respectively.

Perfume Oil or Perfume Compound
A perfume oil is a concentrated mixture of fragrance substances which is used for the perfuming of various products.

Perfumer
Perfumer is the term for the creator of fragrance compositions. His qualifications are: an odor-memory which is the result of long training, the ability to differentiate between hundreds of odors, knowledge of the reciprocal action of individual fragrance substances in the composition, and creative talent. The perfumer's education generally lasts for five years.

Pheromones
Pheromones are chemical substances which make communication possible between living beings. They are mainly significant as a medium which insects use for sexual attraction. Most pheromones are odorless to human beings.

Pomade
Pomade is a substance that is produced through the enfleurage process. In this procedure, animal fat is saturated with blossom fragrance.

The mixture of fat and blossom oil is the pomade. It is either used straight or processed into absolute from pomade. Parts of the fatty odor attach to the fragrance of the blossom oils, giving the products made by this process a unique character.

Powdery
Powdery is the fragrance effect produced by the interaction of long-lasting, mossy, woody, sweet and crystalline elements. Many perfumes leave a powdery overall impression after evaporation of the fresh and floral ingredients.

Rectify
See Distillation.

Resinoids
Resinoids are extracts from resins or plant parts (except for the blossom). In addition to the essential oils, they contain ingredients such as the waxes and resins which are soluble in whatever solvent is being used in the particular process. In order to facilitate the use of resinoids, high-boiling, odorless solvents are often added to them. Resinoids often have a dark color and especially good fixing properties.

Resins
Resins are mainly solid or semi-solid organic plant secretions. They must go through a cleaning procedure before they are used in perfumery.

Round Off
"Rounding-off" means harmonizing and binding together the principal ingredients of a perfume, either with odorants that are closely related odorwise, or with other adjuncts that also fit into the picture and can therefore help to produce a balanced, harmonious whole.

Sensory Adaptation
Sensory Adaptation is the tendency of the human sense of smell to become less and less able to perceive a particular fragrance the longer it is exposed to it. When the sense of smell is "adapted" to a fragrance, it is no longer able to recognize it. Yet it recovers quickly from this fatigue.

Sensual
Sensual is the term for a perfume with an erotically-stimulating effect. An accentuated portion of animalic components and exotic blossom notes is usually to be found in such perfumes. Properly used, many perfumes can produce pleasant emotions and moods, since the sense of smell is directly connected with the part of the brain in which feelings and sexual behavior are controlled.

Smell
Smell is the sensory perception of odorous organic compounds. It occurs in the act of inhaling, and so air is the carrier.

Smoky
Smoky notes are used mainly in masculine perfumes to create natural leather effects. In modern leather notes the smoky notes are thrust into the background by animalic notes but the old, classical leather perfumes contain noticeable smoky notes which originate from birch tar oil.

Solvents
Solvents are liquids, virtually odor- and colorless, used in perfumery for the dilution of perfume oils. The most commonly-used solvent is ethyl alcohol. Some solvents also have fixative properties.

Sour
A perfume is said to smell sour when it has aged prematurely owing to inappropriate storage. When this happens, chemical alterations occur which are irreversible, and the perfume must be considered "off."

Spicy
Spicy fragrance notes are used in the perfume industry in the form of essential oils from almost all the well-known spices. For example, cinnamon and clove are used widely in Oriental perfumes. Many masculine perfumes contain portions of spice-oils–for example: marjoram, coriander and pepper.

Splash Cologne
Splash colognes are light, watery alcohol/perfume oil solutions of 1%–3% perfume oil in 99%–97% alcohol, respectively. They are used generously, for refreshment for the whole body, after the shower or bath for example. They have a subtle perfuming effect, and the notes are fresh and clean. Some countries especially favor this application, and two prime examples are France and Spain.

Stability
Stability is a perfume's resistance to the harmful effects of light and oxygen. In proper storage (protected from light at room temperature, closed bottle) perfumes keep an average of six months without deterioration. Many keep for a much longer period.

Substantivity
The lasting properties of a fragrance are dependent on its degree of volatility. Heavy, nonvolatile substances are used for the fixing of perfume compositions.

Sweet
Sweet elements exist in many perfumes, in differing amounts, especially in Oriental and heavy chypre perfumes. The best-known example of a sweet-smelling natural product is the extract of the vanilla bean.

Tincture
Tinctures are cold-processed alcoholic extracts from natural products. They were much favored in former times, but their use today is on the decline for economic reasons.

Tobacco Notes
Tobacco notes, natural and synthetic, are predominantly used in masculine perfumes. In addition to the pure tobacco note, scents such as honey and plum, which are used to flavor tobacco, also play a part.

Top Note
The top note is the first phase of a perfume's evaporation "life." It is a decisive part of the first impression, on opening the bottle and application of the perfume to the skin. The function of the top note is to awaken interest for the perfume in general, and to provide impact. Therefore, for a top note, a striking character is often more important than complete harmony. The top note is ruled by light, volatile fragrance materials, but often, notes from heart and foundation also play a part in the first fragrance phase.

Trainee
Trainee is the word for the student perfumer in the first two to three years of his perfume-education. During this phase of his career, he is mainly involved in the training of his sense of smell; and in developing an odor-memory. The whole education takes approximately five years, and takes place in the laboratories of the fragrance and perfume industry.

Vapor Pressure
Vapor pressure is the change of liquid molecules into molecules of gas. The thinner and more volatile the material, the higher the vapor pressure. Vapor pressure is affected by varying temperatures.

Volatility
The volatility of fragrance substances used in perfumery depends, in the case of single chemical entities, on the size of the molecules. In natural substances which are mixtures, it is dependent on the diverse volatilities of the components. Highly volatile substances determine the first olfactory impression of a perfume; less volatile substances, the after-effect. Since perfume is a complex mixture of substances of differing volatilities, changes in character during the fragrance evaporation are normal. It is an important part of the perfumer's art to give the perfume a uniform, basic character during every phase of the evaporation.

Warm
Warm perfumes have a high portion of ingredients with animalic fragrance character, and make a somatic impact.

Woody
Woody notes appear, in varying degrees, in almost every perfume. The most important natural woody oils in perfumery are cedarwood oil, patchouly oil, vetiver oil and sandalwood oil. They all have good fixing qualities, consequently their characteristic effect develops only in the later stages of fragrance evaporation. They are responsible for the powdery after-effect left on the skin by many perfumes. Many modern men's perfumes owe their character to synthetic woody fragrance substances developed in the last few years by the fragrance material industry.

The Production of Fragrance Materials: from Nature and Chemistry

The Production of Natural Fragrance Components.

Living Nature, with her great abundance of flora, offers an enormous reservoir of fragrance materials of vegetable origin. The animal kingdom, however, only supplies four odorants that are of any importance to the perfume industry.

Although it is the blossom that usually makes the most striking and characteristic contribution to a plant's odor-complex all other parts can also be made to yield fragrant material. For example: blossoms (rose, jasmine, tuberose, etc.); stems and leaves (geranium, patchouly, peppermint, etc.); fruits (anise, coriander, nutmeg, etc.); seeds (celery, cardamom, parsley, etc.); fruit peels (orange, lemon, bergamot, etc.); roots (angelica, orris, vetiver, etc.); wood (sandalwood, cedarwood, rosewood, etc.); herbs and grasses (tarragon, sage, thyme, etc.); needles and twigs (spruce, Scotch pine, cypress, etc.); resins and balsams (galbanum, myrrh, olibanum, etc.); bark (cinnamon, etc.).

Some plants are multi-producers. For instance, the bitter orange tree. Its blossoms yield either orange flower absolute or neroli oil, its leaves petitgrain oil, the peel of its fruit orange oil.

Though the contribution of the animal kingdom is small, it is invaluable.

Amber, also known as ambergris, presumably a pathological secretion of the sperm whale, is either found floating in the ocean, or taken from the body of the whale.

Musk is a sexual secretion of the musk deer.

Castoreum is a sexual secretion of the beaver.

Civet is a secretion of the civet cat, found in both sexes of this animal.

Castoreum can be obtained only by killing the animal, and so it is with musk–though genuine musk has all but disappeared from the fragrance palette of the perfumers. Most of the amber used in perfumery is gathered from the sea. Civet is obtained without harming the animal, and it is taken from cats especially bred for the purpose.

From the Rose to the Flower Oil.

The easiest way to enjoy Nature's fragrances is to bring cut flowers into your home or to grow them in pots. Our grandmothers sewed dried lavender into little fragrance sachets and put them into the laundry closets. During Advent, the Christmassy smell of pine twigs fills many homes.

This is about all an amateur can do to preserve Nature's fragrances.

The very fragility and evanescence of fragrances have always made them all the more desirable. Therefore, man has always looked for ways to isolate the beautiful scents of Nature, learn their composition, and preserve them. Today, we know that they are basically essential oils –volatile, oily substances enclosed mainly in cellular glands of the leaf and blossom surfaces, but also in secretion cells deep inside other parts of the plant. The oldest method of extracting the flavor and fragrance materials of plants was boiling and brewing–the way we prepare a tea or herb brew today. The most common methods in use today for obtaining natural fragrance components are based on the old principles of steam distillation, expression and extraction but are more efficient, technically.

Distillation: Fragrance in Steam.

Distillation is still the most economical method of extracting essential oils from plant materials.

Before the distillation begins, the material is broken up in order to open the oil cells and make them more accessible to the steam. Blossoms and leaves, in which the oils occur in thin-walled cells directly under the surface, need no further preparation. Roots and herbs are cut up, dried plants are ground, seeds are crushed, woods are grated. These processes naturally do not open up all the cells–part of the oil extracted will come from closed cells. However, the better the preparation, the faster the distillation.

Florentine flasks
for oils which are lighter than water
for oils which are heavier than water

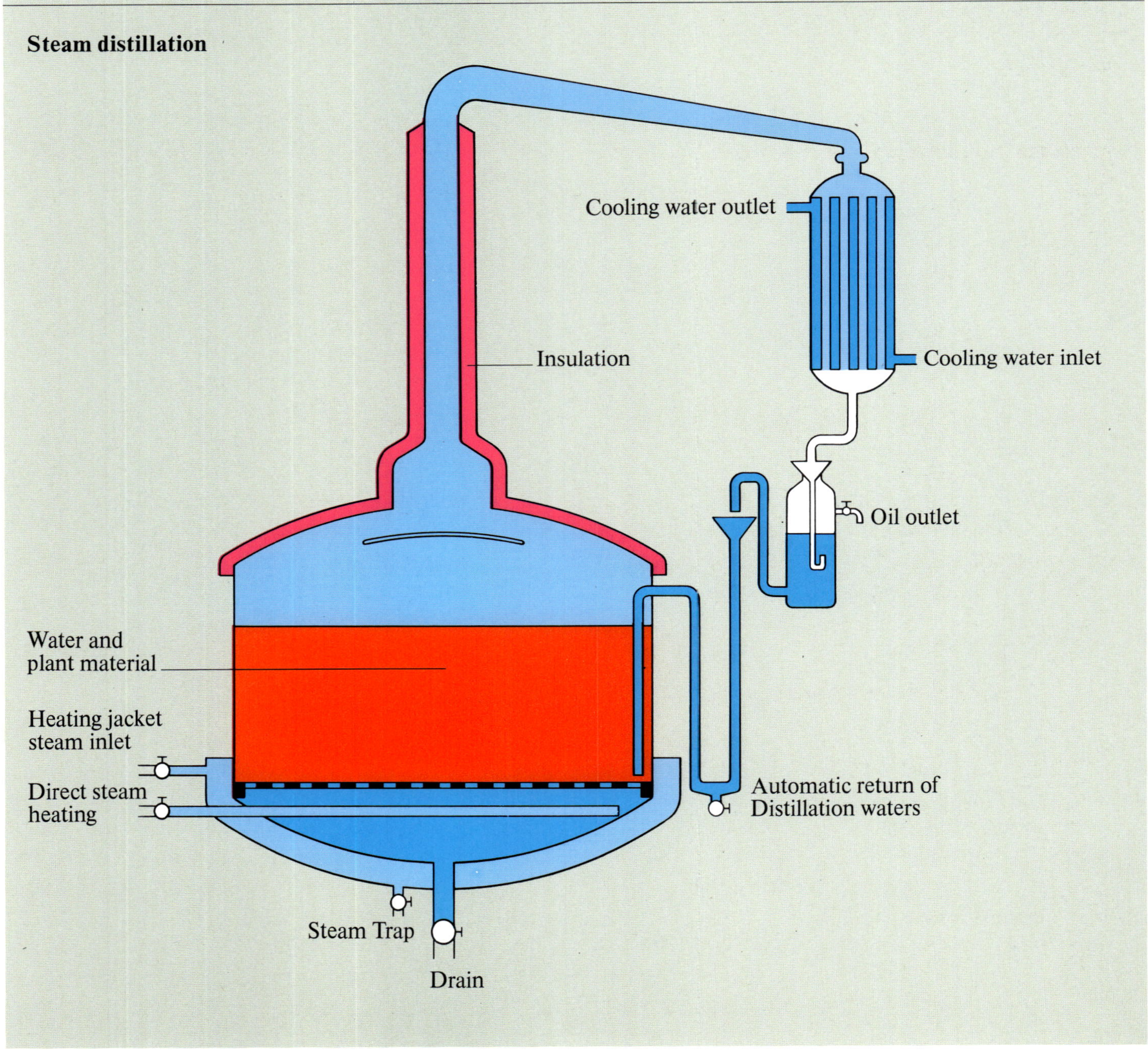
Steam distillation
Cooling water outlet
Insulation
Cooling water inlet
Oil outlet
Water and plant material
Heating jacket steam inlet
Direct steam heating
Automatic return of Distillation waters
Steam Trap
Drain

Vacuum destillation unit
for the removal of traces of solvents from natural flower oils

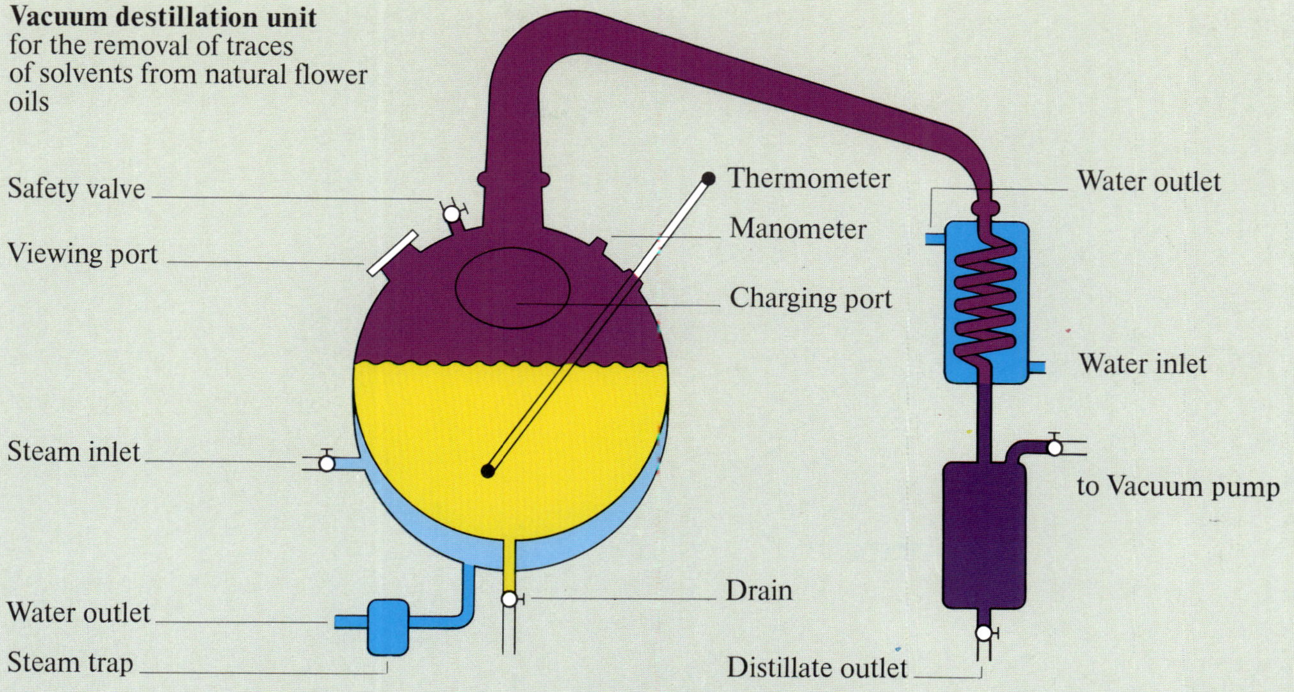

Extraction

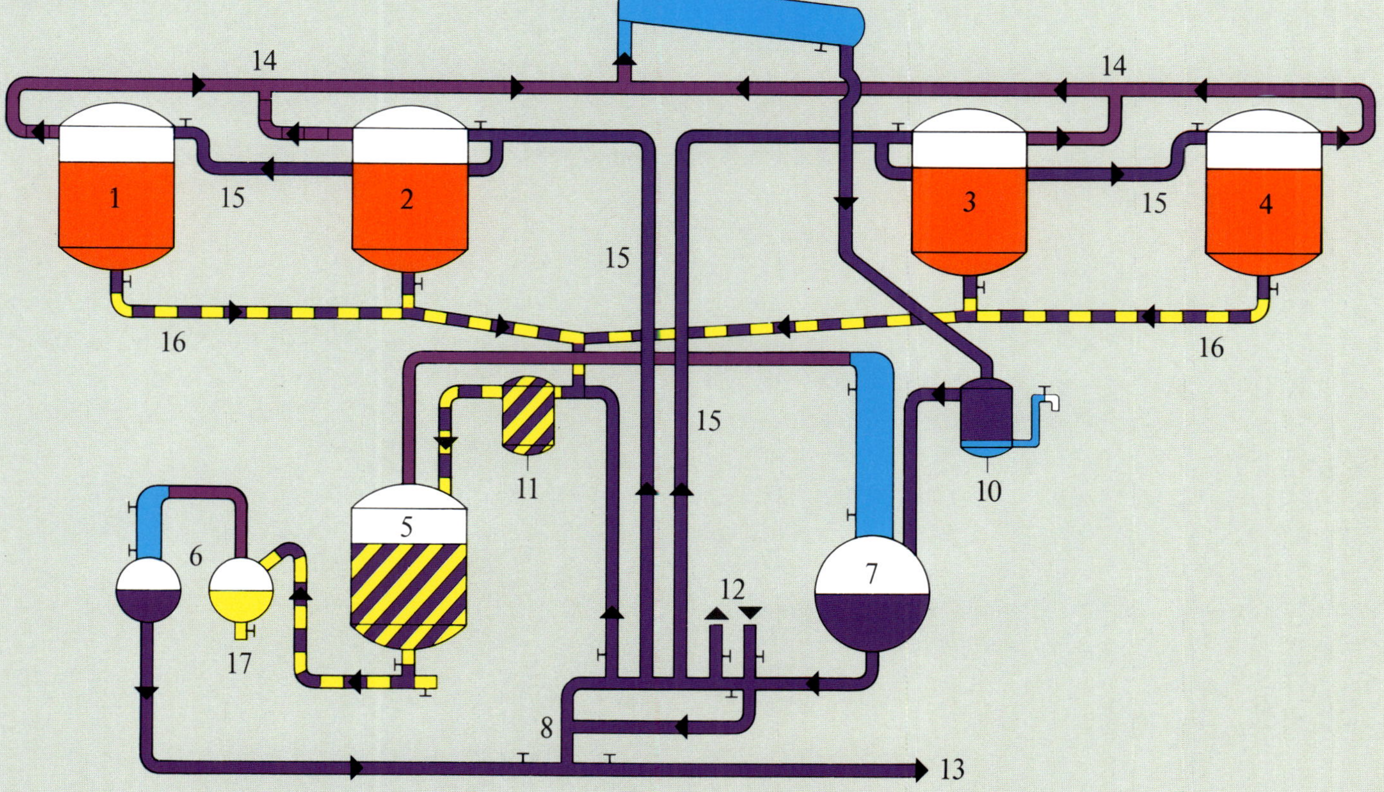

1 · 2 · 3 · 4 Extractors.
5 Evaporator – for the removal of most of the solvent.
6 Vacuum distillation equipment for the removal of the last traces of solvent.
7 Distillate receiver (solvent).
8 Solvent pump.
9 Condenser for solvent vapours.
10 Settling tank for solvent.
11 Settling tank for extracts prior to solvent removal.
12 To and from solvent tanks.
13 To solvent rectification.
14 Overflow for solvent vapours from the extractors.
15 Solvent flow to extractors.
16 Extract flow from extractors to the evaporators.
17 Outlet for Solvent – free extracts (concrète).

yellow	concrète
violet	solvent
red	material to be extracted
blue	water

The most simple form of distillation, water distillation, is no longer used. In this procedure, the distillation materials were completely immersed in water and boiled. This method was particularly suitable for delicate oils such as rose and neroli (orange blossom) oil.

In the steam distillation method, the distillation materials are placed on a grille, and super-heated steam is directed through them. The intense heat of the steam is able to liberate even fragrance materials with extremely high boiling points.

Extraction with Animal Fats.

In addition to distillation, the procedure of fat extraction, known as *Enfleurage,* has been in use for centuries for obtaining blossom fragrances. Today, this is already perfume history.

The method was as follows: a sheet of plate glass was placed in a wooden frame (the "chassis"), and spread with lard into which the flower petals were pressed. Within a few days, the fat absorbed the fragrance from the plant material.

The same fat was used over and over again, until it was saturated. The resulting grease/fragrance mixture was called pommade. The fragrance materials were then extracted with alcohol and the result was known as absolute from pommade *(essences absolues de pommade).* This was an extraordinarily wasteful and costly procedure, and is no longer used today.
For the creative work of the perfumer, however, these absolutes were very interesting. They differ very considerably in fragrance characteristics from materials extracted by volatile solvents. The reason is that during the extraction process, an enzymatic decomposition takes place which changes the odor and gives higher yields.

Maceration, another method no longer in use, is extraction with warm fat. It is also known as *enfleurage à chaud.*

Extraction with Volatile Solvents.

The first recorded experiments with this method took place around the beginning of the last century. Mass production began shortly before the beginning of this one. The big Grasse companies have been especially involved in this development.

The solvents which are used must have certain important characteristics: they must be pure, leaving no residue after evaporation, and they must have a low boiling point so that fragrance components are not destroyed by the heat. The boiling point should not be too low, either, or there will be excessive losses due to evaporation. Finally, the solvent must be selective, only extracting the substances desired. A preferred solvent is petroleum ether. Recently, carbon dioxide has become significant. Because of its low boiling point, it is the least destructive of the known extraction solvents. It was carbon dioxide that made it possible to extract the fragrances of lily of the valley and lilac. These two, among the most prominent and unique scents in Nature, are so delicate that all other extraction processes had failed.

In the solvent method of extraction, the raw material is placed on grilles in closed containers, and the solvent flows through them. Alternatively, the containers, with the raw material inside, can rotate through the solvent.

A wide variety of substances may be extracted by this method: from fresh plant parts such as blossoms and leaves come waxes and color substances in addition to fragrance materials. After the evaporation of the solvent, there remains a wax-like, relatively firm mass which is called the concrete *(essence concrète).* After further treatment to remove the waxes from this substance, we have what is known as the absolute *(essence absolue).* From the extraction of plant resins we derive the substances that are known as resinoids.

The odorants of animal origin are processed to *tinctures* and *infusions.* These are alcoholic extracts of amber, musk, castoreum and civet.

The citrus oils of fruits such as orange, lemon and bergamot are extracted by *expressing,* pressing the peel.

Flowers from the Retort.

No naturally occuring scent goes straight into the perfume bottle. All the processes of fragrance extraction are complex. The supply of raw materials is also difficult. Land available for cultivation is becoming more and more limited as the world's population grows. The plundering of the tropical forest, the decline of natural flora through agricultural monocultures, other influences of encroaching civilization, and hunting have considerably reduced the world's supply of fragrance raw materials.

Even though the civet cat is now especially bred and fragrance plants have been cultivated for centuries, the natural raw materials available would cover only a third of the world's needs. As a result, synthetic fragrance materials have become more and more important. The synthesis of fragrance components is not restricted to the imitation of natural odors. Substances have been created in the laboratory which do not exist in Nature at all. Among these are well-known fragrance components such as hydroxycitronellal, which has a beautiful, floral effect reminiscent of lily of the valley and lilac.

Furthermore, certain perfume effects, which could never have been achieved with natural fragrance materials, are possible with synthetics. These have shaped complete fragrance directions, such as the aldehyde notes (Chanel N°5) with the fatty aldehydes; or the floral notes, with benzyl salicylate (L'air du Temps). Two further examples: light, modern perfumes such as Diorella could not have been created without methyl dihydro jasmonate; and chypre perfumes such as Cabochard could not have been created without isobutyl quinoline.

The Production of Synthetic Fragrance Components.

The use of fragrant materials from plants and the animal kingdom, first for religious purposes, but later, increasingly for personal use, dates back to antiquity.

Compared to that, the history of synthetic fragrance substances is still very short. It began in the first half of the last century and ran parallel to the rapid development of organic chemistry. Even in those days, it was a fascinating job for the research chemists to trace the "fragrance principles" of natural perfume raw materials–in other words, to isolate them from the more or less complex natural mixtures in which they occur, and to elucidate their chemical structures.

Here are a few milestones in the development of synthetic fragrance materials:

1834 Isolation of cinnamic aldehyde from cinnamon oil (Dumas, Péligot)

1837 Isolation of benzaldehyde from bitter almond oil (Liebig, Wöhler)

1853 Synthesis of aliphatic aldehydes (Piria)

1856 Synthesis of cinnamic aldehyde (Chiozza)

1863 Synthesis of benzaldehyde (Cahours)

1874 Synthesis of vanillin from coniferin (Haarmann, Tiemann)

1875 Synthesis of coumarin (Perkin)

1876 Synthesis of vanillin from eugenol or guaiacol (Reimer, Tiemann)

1893 Synthesis of ionone (Krüger, Tiemann)

1904 Isolation of muscone from glands of the musk deer (Walbaum)

1926 Identification of the muscone structure (Ruzicka)

1934 Synthesis of muscone (Weber, Ziegler)

The systematic investigation of odor-determining main ingredients of natural fragrance materials has in many cases initiated the industrial production of important synthetic fragrance materials. In recent times, the interest of the fragrance chemist has been directed increasingly towards the fragrance-intensive by-products and trace-components which–even in small concentrations–produce a characteristic effect and are therefore especially valuable to the perfumer.

Classification of Aroma Chemicals.

Aroma chemicals can be produced from natural as well as synthetic raw materials, and are divided into the following groups:

Isolates. These are obtained from essential oils or other natural materials, with the help of physical or chemical separation methods, for example, eugenol from clove leaf oil, eucalyptol from eucalyptus oil, cedrol from cedarwood oil, citral from lemongrass oil, and menthol from peppermint oil. Eugenol, eucalyptol and cedrol are only available as isolates. Citral and l-menthol are also synthetically produced.

Semi-synthetic Aroma Chemicals. These are produced from isolates through chemical reactions, for example, l-carvone from limonene, cedryl acetate from cedrol, hydroxy citronellal from citronellal, or terpineol from pinene.

Synthetic Aroma Chemicals. These are synthesized from basic organic chemicals (from coal or petroleum). These are predominantly nature-identical aroma chemicals, that is, synthetic chemicals modeled after naturals, with the identical chemical structure, or, on a lesser scale, chemicals not found in Nature, new structures that resemble natural odor types nevertheless.

Arranging aroma chemicals by their chemical structures, gives the following general groupings: *Terpenes.* They are considered a special group due to their importance and close relationship, although strictly speaking, they belong to the aliphatic or alicyclic aroma chemicals. Examples: menthol, geraniol, citral. (Fig. 1)

Aromatics. Also known as benzenoids, the most important of these aroma chemicals are produced from benzene, toluene or phenol. They are the largest group, in terms of production as well as revenue.

Aliphatics. Of interest are the fatty aldehydes, some alcohols and fatty acid esters. Altogether, this group is fairly small.

Alicyclics. This group contains some very important jasmine chemicals such as cis-jasmone and dihydro methyl jasmonate, as well as the macrocyclic ketones muscone and civettone.

Heterocyclics. The lactones have special significance (gamma-undecalactone, cyclopentadecanolide, etc.). Heterocyclics with nitrogen or sulphur (pyridine, pyrazine, thiophene, thiazole, etc.) are trace components with strong odor and are only used in small amounts.

Industrial Production of Synthetic Aroma Chemicals

Using terpene-products and aromatics as examples, the industrial production of important odorants from relatively few basic raw materials is illustrated by graphs, or "family trees"–see illustrations 1–7.

Terpene Compounds. (Fig. 1) For a long time, essential oils were the only available source of terpene aroma chemicals, which accordingly were isolates or semi-synthetic aroma chemicals. Ionones and methyl ionones from citral (ex-lemongrass or litsea cubeba oils) or hydroxycitronellal and l-menthol from d-citronellal (ex-citronella oil) are well known examples of these partial syntheses.

Since the 1960's, the natural sources could no longer cover the increasing world-wide demand for aroma chemicals for use in cosmetics and household products. At the same time, due to the synthetic production of vitamins A and E, an additional demand arose for citral and linalool, which hitherto were only available from essential oils.
The large vitamin producers solved the raw materials problem through the total synthesis of the terpene key ingredients from simple organic chemicals (acetone, acetylene, isoprene). The important intermediate products are methyl heptenone and dehydro linalool.

Partial syntheses are possible, starting from alpha or beta pinene obtained in abundant quantities by fractional distillation of turpentine oil (the low-boiling fraction of gum turpentine from conifers).

Myrcene is obtained from beta pinene by pyrolysis, and is converted by various processes to linalool or geraniol/nerol. The subsequent manufacture of products like citronellol, citral, ionone, etc., is carried out in the same manner as when starting materials were natural isolates.

For a long time, alpha pinene served in the manufacture of terpineol and isobornyl acetate, but only recently the industrial syntheses of linalool and geraniol/nerol from alpha pinene or alpha/beta pinene mixtures have been achieved. The process runs via cis-pinane, pinane hydroperoxide and pinanol to linalool, which ultimately is isomerized (rearranged) to geraniol/nerol.

Aromatics. (Figs. 2–7)

All aromatics have the benzene-ring in common and could be shown as one family. However, this would be too confusing, and would not do justice to certain raw materials and structural groupings. The necessary raw materials are either directly obtained from coal or petroleum (benzene, toluene, xylene, naphtalene) or through chemically treating benzene or toluene (phenol, cresol).

From benzaldehyde, the main intermediate derivative of *toluene,* a large number of important aroma chemicals are synthesized, for example, cinnamic aldehyde, alpha amyl and alpha hexyl cinnamic aldehyde. Alkylation of toluene with isobutylene produces para tertiary butyl toluene, the starting material for lilial.

Phenyl ethyl alcohol is produced from *benzene,* and via cumol as an intermediate product, cyclamen aldehyde and galaxolide are produced.

Among the *cresols,* meta cresol has special significance for the syntheses of menthol (via thymol) and musk ambrette.

The *phenolics* represent the largest group of aroma chemicals, both tonnage-wise and in terms of revenue. Important aroma chemicals derived from phenol include: coumarin via salicylic aldehyde, anethol via anisole, and salicylates from salicylic acid. Pyro-catechol, that can also be obtained from phenol is, among others, the basic raw material for synthetic vanillin, isoeugenol and sandal.
Figures 1 through 7 demonstrate products that represent money-wise approximately 80 % of all synthetic aroma chemicals. Their industrial significance will increase in the coming decades. This positive outlook results from the fact that synthetically produced aroma chemicals–compared to natural–are available in consistent qualities, at relatively constant prices, and in practically any quantity desired.

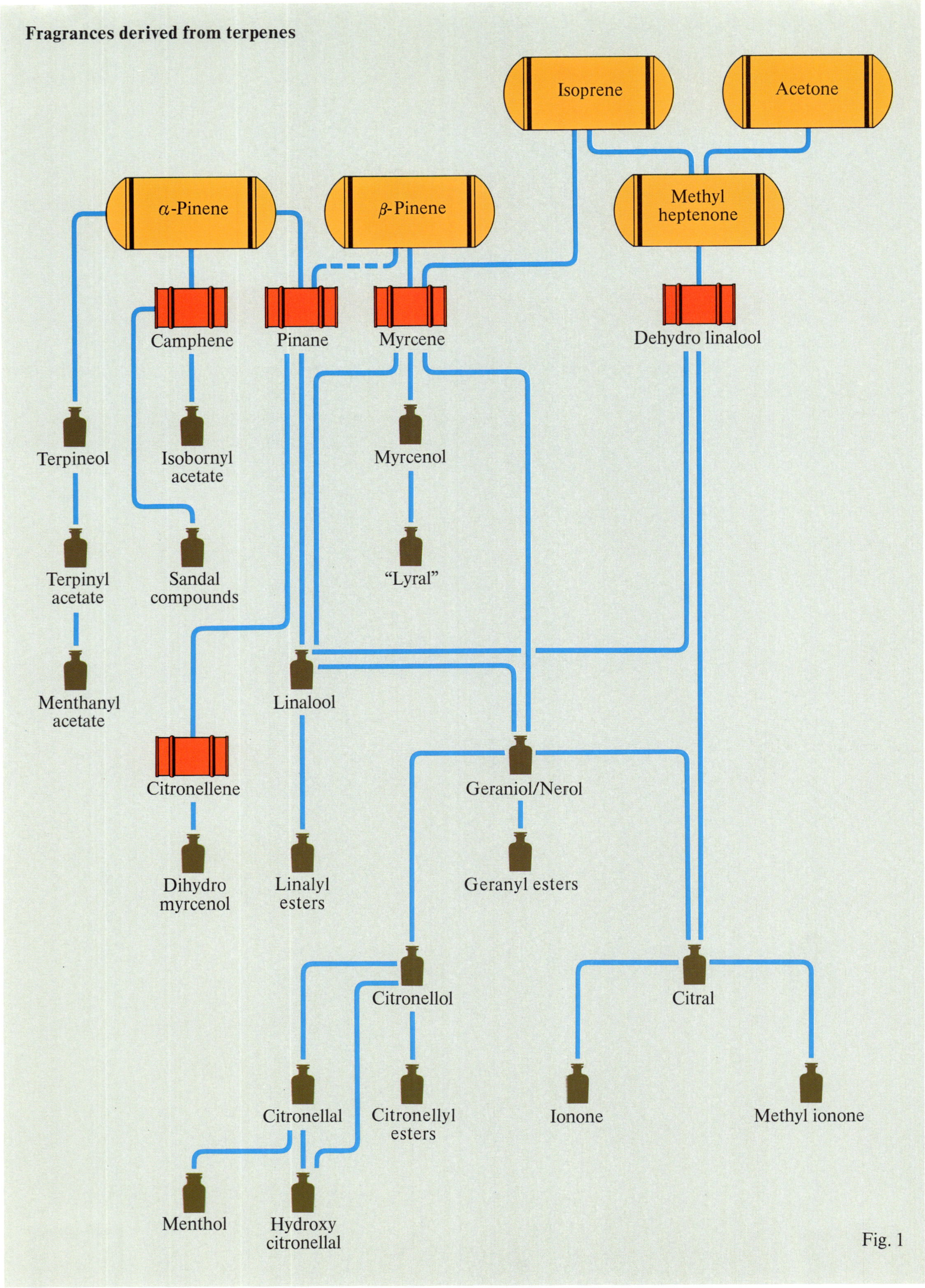
Fragrances derived from terpenes
Isoprene
Acetone
α-Pinene
β-Pinene
Methyl heptenone
Camphene
Pinane
Myrcene
Dehydro linalool
Terpineol
Isobornyl acetate
Myrcenol
Terpinyl acetate
Sandal compounds
"Lyral"
Menthanyl acetate
Linalool
Citronellene
Geraniol/Nerol
Dihydro myrcenol
Linalyl esters
Geranyl esters
Citronellol
Citral
Citronellal
Citronellyl esters
Ionone
Methyl ionone
Menthol
Hydroxy citronellal

Fig. 1

Fragrances derived from toluene

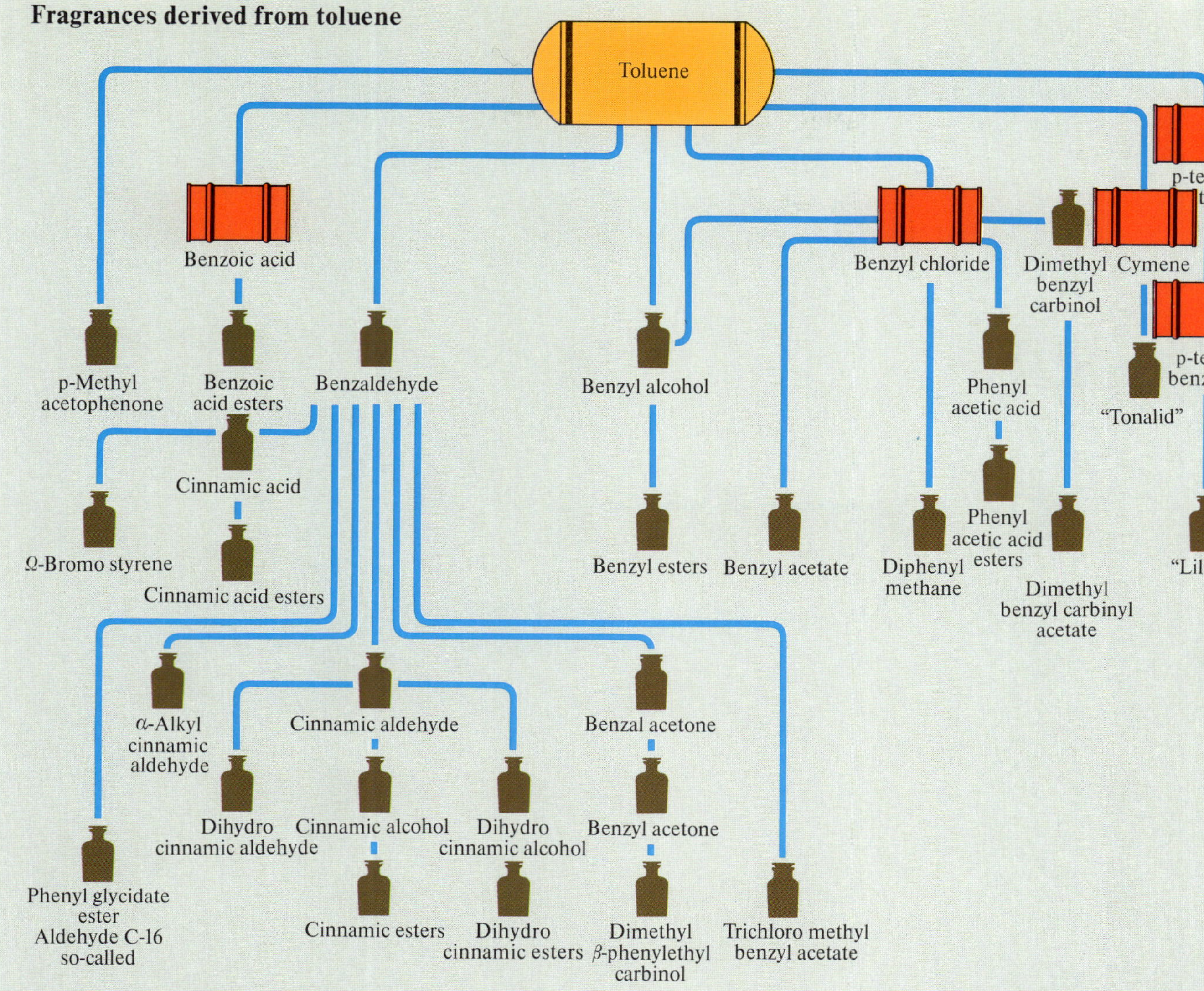

Fig. 2

Fragrances derived from Naphthalene

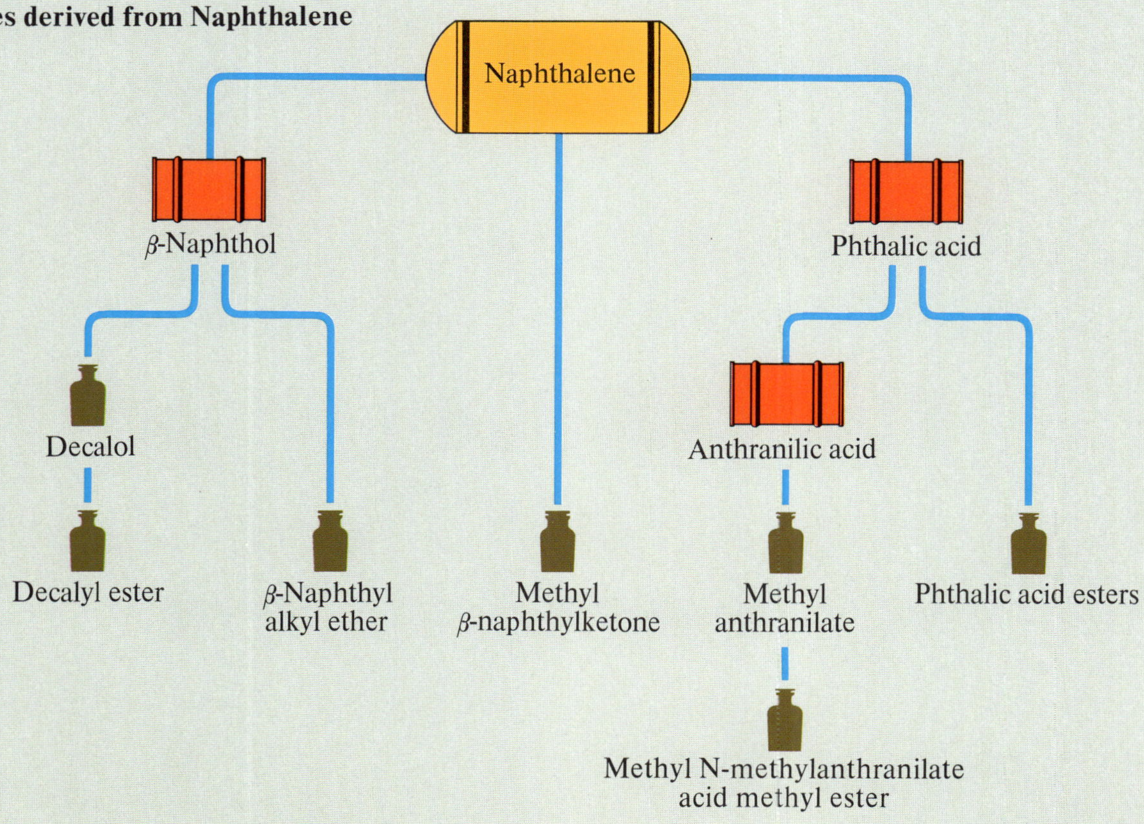

Fig. 3

Fragrances derived from benzene and m-xylene

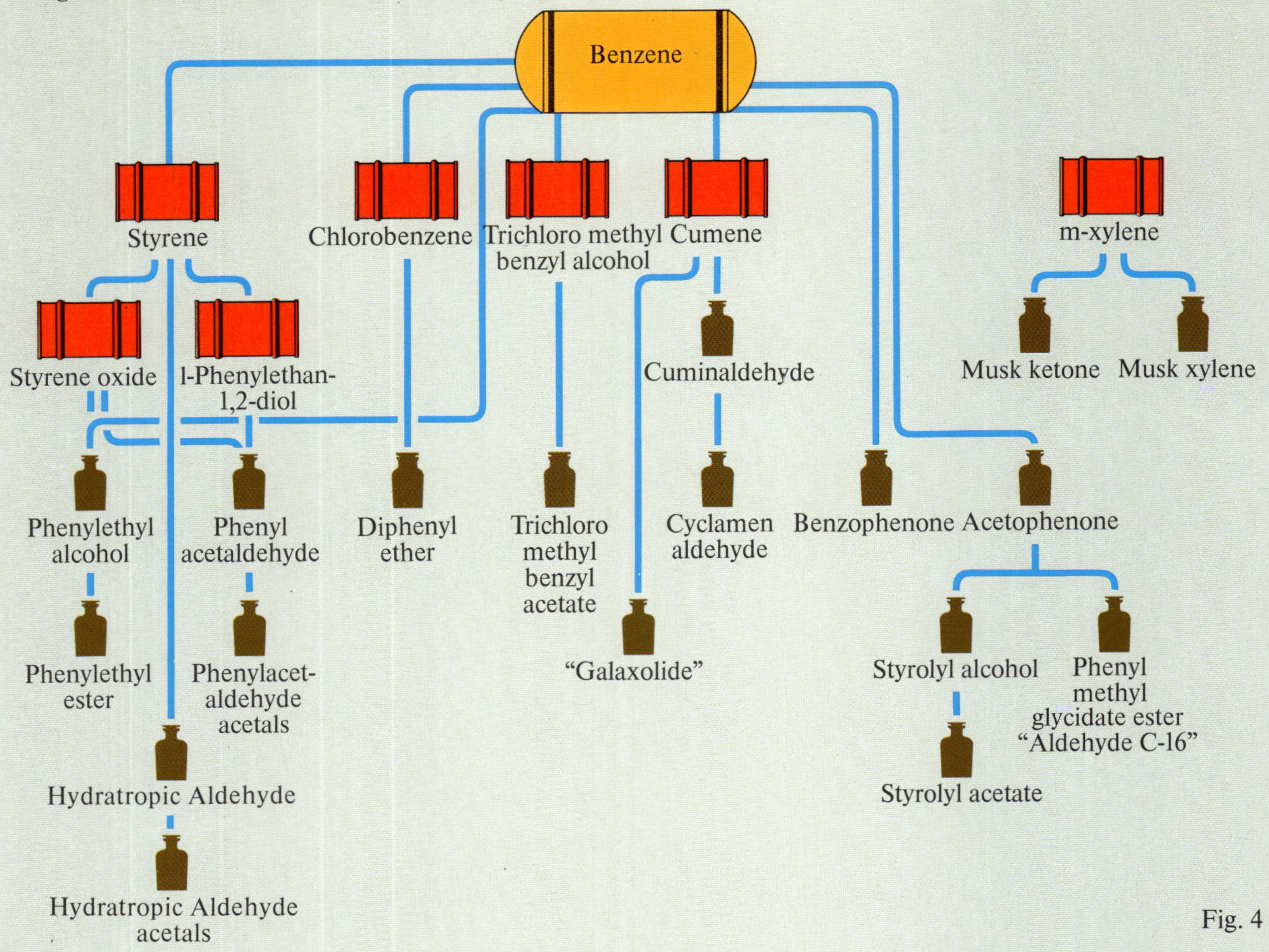

Fig. 4

Fragrances derived from the cresols

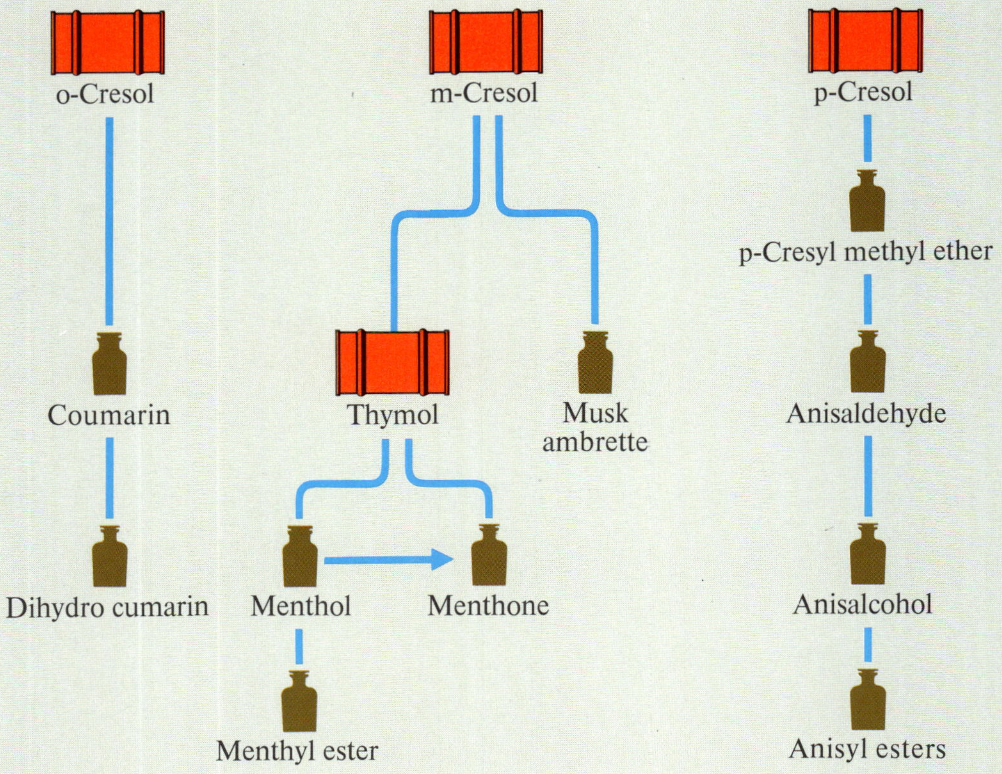

Fig. 5

Fragrances derived from phenol but not including pyrocatechol, resorcinol and hydroquinone derivatives.

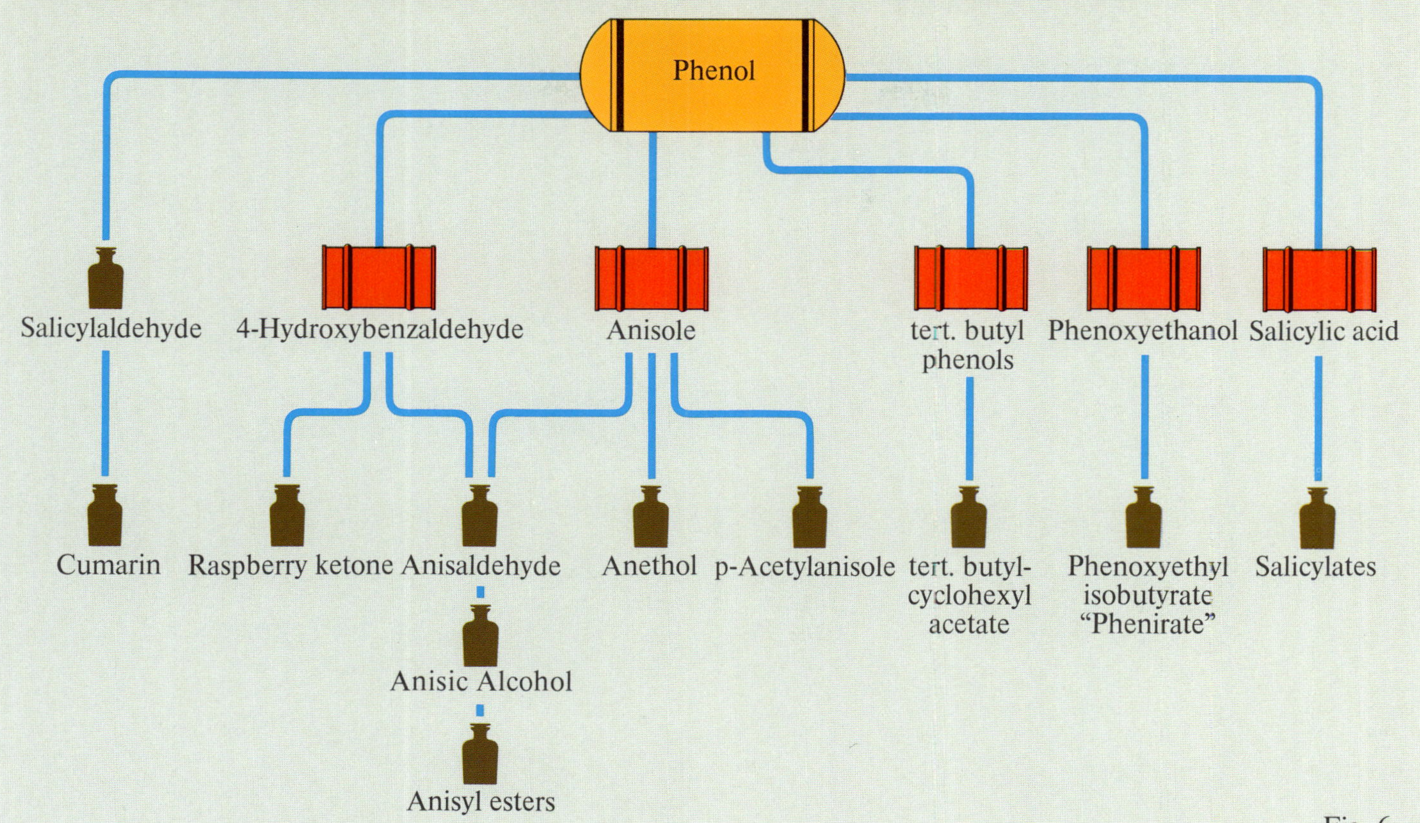

Fig. 6

Fragrances derived from pyrocatechol, resorcinol and hydroquinone

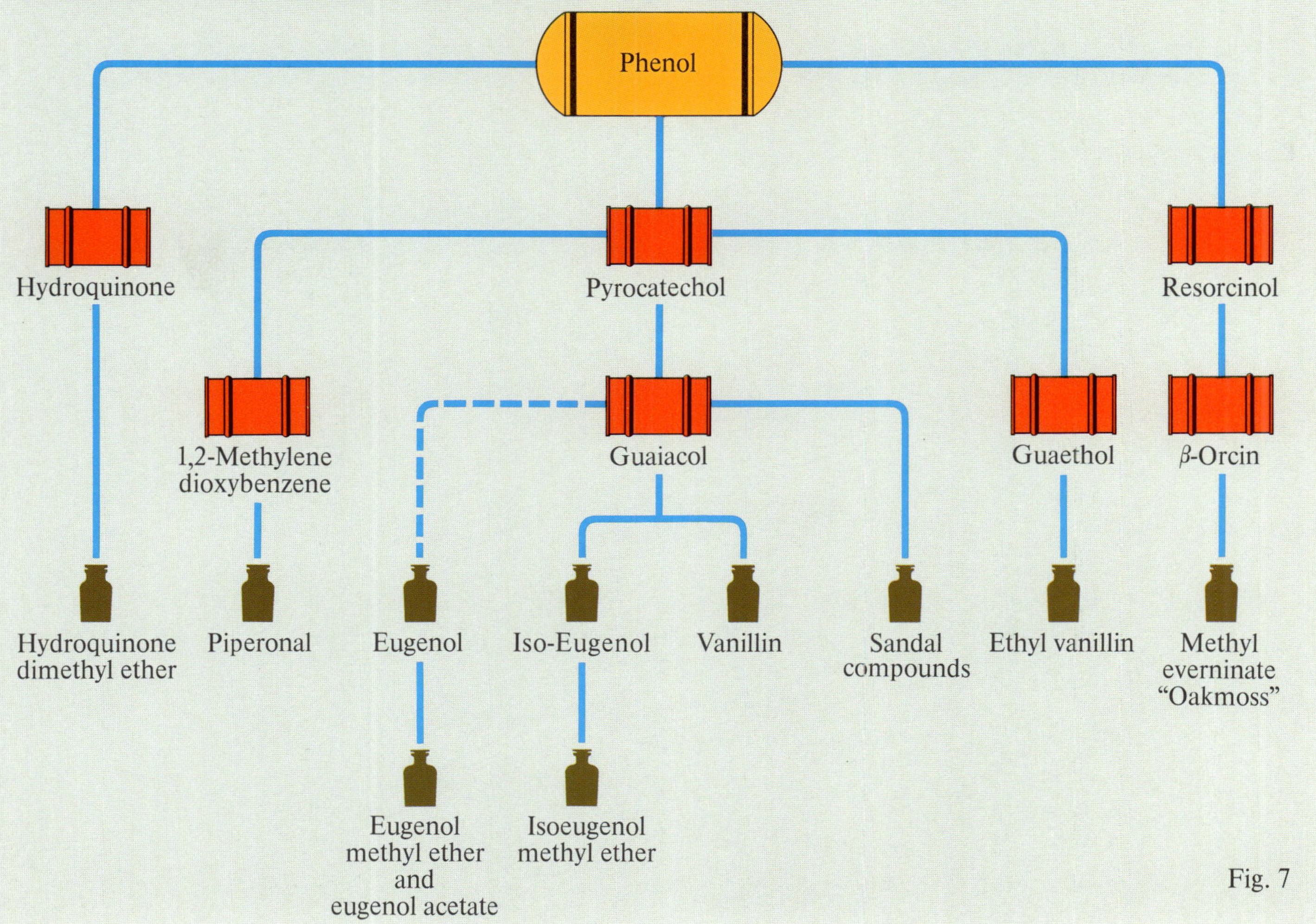

Fig. 7

What Comes from Where?

A Selection of the most important Natural Fragrance Components and the Geographical Origins of their Raw Materials.

Angelica root oil, Root	1
Benzoin Siam resinoid, Resin	2
Bergamot oil, Peels	3
Cananga oil, Blossoms of the Ylang Ylang tree	4
Castoreum, canadian and siberian beaver	5
Cinnamon bark oil, Bark	6
Cistus oil, Leaves and twigs	7
Citronella oil java, Grass	8
Civet absolute, Civet cat	9
Clove oil, Clove	10
Cypress oil, Leaves and twigs	11
Eucalyptus globulus oil, Leaves and twigs	12
Fir needle oil siberian, Twigs	13
Galbanum resinoid, Gum resin	14
Geranium oil, Leaves and stalks	15
Jasmin absolute, Blossoms	16
Lavandin oil, Herb	17
Lavander oil, Herb	18
Lemon oil, Peels	19
Mandarin oil, Peels	20
Musk tincture, Musk-deer	21

Neroli oil, Blossoms of the bitter orange tree 22
Oakmoss resinoid, Oakmoss 23
Olibanum resinoid, Resin 24
Orange oil sweet, Peels 25
Orris oil, Root 26
Patchouli oil, Leaves 27
Peppermint oil, Herb 28
Petitgrain oil, Leaves and twigs of the bitter orange tree 29
Rose oil/-absolute, Blossoms 30
Rosemary oil, Herb 31
Sage oil, Herb 32
Sandalwood oil, Wood 33
Spearmint oil, Herb 34
Spike Lavender oil, Herb 35
Vetiver oil, Root 36
Ylang Ylang oil, Blossoms 37

Index →

Alphabetical Index

Lavender: Cultivation and Production of the Essential Oil

Lavender is probably the most-used essential oil in the perfume industry. It used to be extracted from wild plants in the Mediterranean Region; but today it is cultivated in huge fields in France, Spain, Italy, Morocco and, above all, Yugoslavia.

The plants are processed in their entirety, but only after they have withered. Some farmers still harvest their lavender crops with sickles, but most lavender-harvesting today is done by machine. After the cutting, the plants are dried and bundled, and distilled by steam. When the oil and water mixture has been extracted the spent plants

are dried and used as fuel for the still. One ton of plants yields between eight and ten kilograms of essential oil.

Southern France is known as the classical area for the cultivation of lavender–"Lavandula Officinalis"–but there is another important lavender land: Yugoslavia. Cultivation began here in the year 1910, with 10000 seedlings, in areas where the climate and soil conditions were similar to those in the south of France. The Yugoslavian lavender fields are owned by the farmers, the distillation facilities belong to the community, and each planter must contribute his share of the work.

In perfumery, lavender is used to give "clean" and "fresh" impressions. Since ancient times, the plant has been used for medicinal purposes. The Persians, Greeks and Romans burned lavender twigs, which gave off a highly-odorous smoke, as a means of defense against epidemics and "evil humors" in sick rooms. In the northern part of Europe, the smoke from mixtures of lavender and other plants was used to purify houses in which there had been victims of the plague. The brew of

lavender was also a standard ingredient of health drinks. Eaux de Cologne in which lavender provides a part of the fragrance have not only been used for cosmetic and hygenic purposes, but also taken as medicines. The lavender sachets which our grandmothers used to freshen the closets and the linen shelves were thought to have a certain disinfectant action also. It wasn't until the end of the Nineteenth Century that fragrance came to be considered the main purpose of this much-used plant. At present, lavender is still very much a "clean" connotation, frequently used in soaps, bath products and refreshing fragrances.

Due to widespread cultivation, mechanical handling and modern distillation methods, lavender oil is still a relatively inexpensive raw material. The use of lavender water as a scent has been somewhat out of vogue since World War II, but the oil still has an important role as a nuanceur. It is to be found mostly in Nature-oriented compositions, in the Fougère and Chypre types, and in many other fragrance directions. As a "fragrance component" lavender is still widely used but discreetly, in the background.

The Bitter-Orange Tree: One Plant, Three Raw Materials, Many Fragrance Components

Mention the Scents of Nature, and everybody thinks immediately and automatically of flowers in bloom. But in perfumery, natural fragrance substances from every part of the plant are used: Leaves, grasses, needles, twigs, barks, fruits, seeds, roots and wood–and, of course, the flowers. Individual plants are usually "specialists", supplying perfume ingredients only from one part of their make-up–the blossoms, or the leaves, or what have you. The bitter orange tree is an exception. It supplies the perfume makers with no less than three different and very important fragrance components: neroli oil from the

blossoms, bitter orange oil from the peel of the fruit, and petitgrain oil bigarade from the leaves and twigs.

On our chart (p. 94) the production of neroli oil is shown on the left: The oil is extracted from the blossoms of the bitter orange tree by steam distillation. One thousand kilograms of the hand-picked blossoms produce one kilogram of oil. A byproduct of the steam distillation is the condensate known as "orange blossom water,"–Eau de Fleurs d'Oranger. An extract of this product produces the Absolue de L'Eau de Fleurs d'Oranger.

When the fragrance substance is extracted from the blossoms with a solvent, the first result is orange flower concrete, which through a further extraction process becomes orange flower absolute.

Neroli oil and orange flower absolute are two fragrance components which have little family resemblance; their fragrances are very different.

The main producer-countries are Morocco, Tunisia, Algeria and Egypt.

The central stem on the chart indicates that the fruit supplies the essential oil. Orange oil is extracted from the peel of the fruit. The food industry uses both the peel and the fruit but that is a different matter.

The right-hand column of our chart deals with the green parts of the tree: Leaves, twigs and unripe fruit. From these, Petitgrain Oil Bigarade is obtained by steam distillation. A by-product of this process is Eau de Brouts from which Absolue Eau de Brouts can be extracted.

Petitgrain oil–not only the petitgrain oil from the bitter orange tree, but from other sources, too–is extremely important to the perfume industry. It takes approximately 250 kilograms of raw material to produce one kilogram of essential oil. The annual world production is between 320 and 370 tons.

Fruit, Flowers, Leaves – three essential oils

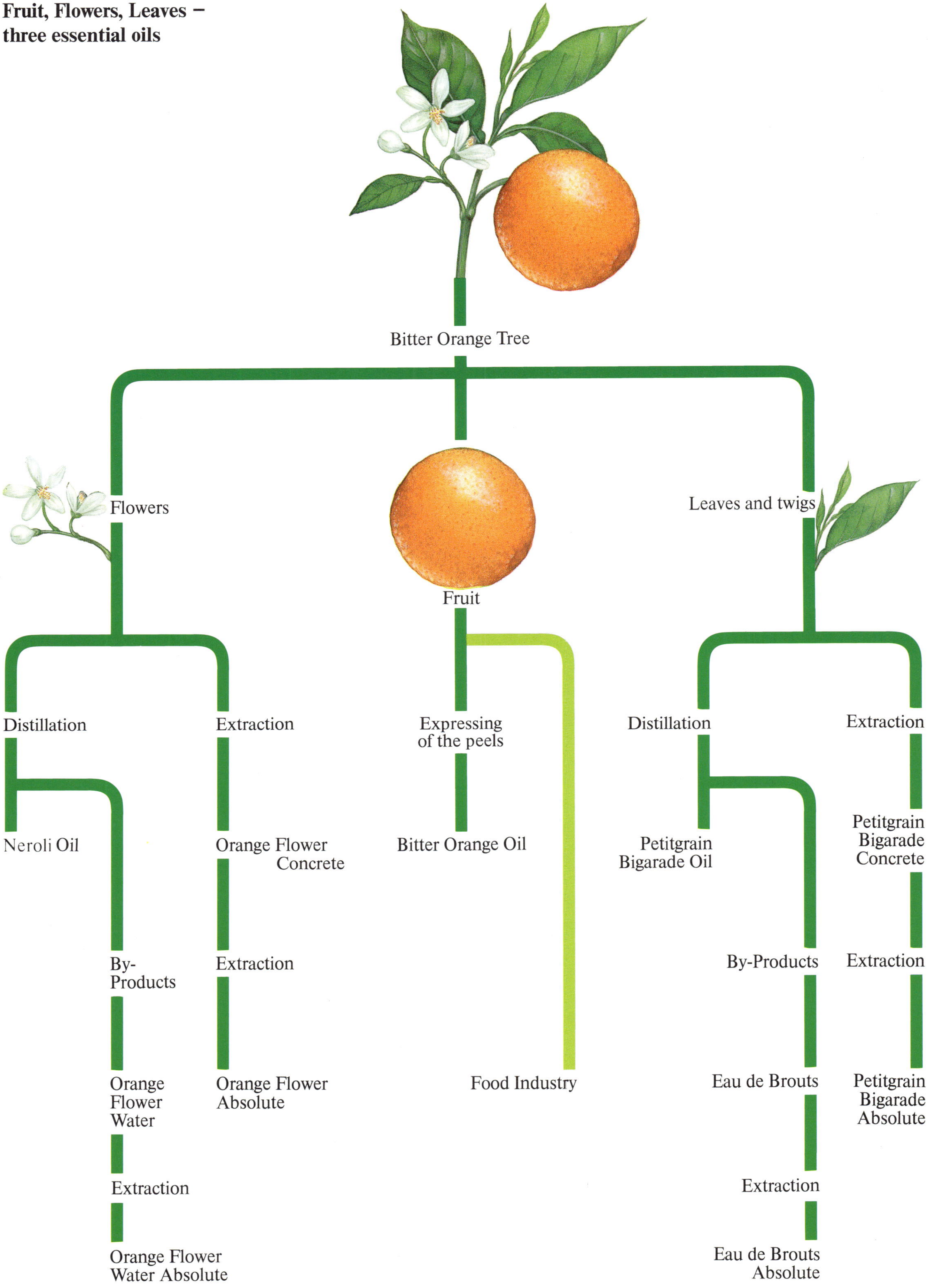

Extraction–as opposed to distillation–of the leaves and twigs first yields Petitgrain concrete Bigaradier which can be processed to the absolute. This latter odorant has a peculiar, unusual note and plays a minor part only in present-day perfumery.

So far as neroli, orange and petitgrain oils are concerned, the bitter orange tree does not deliver all its different fragrance substances at the same time. The reason is that the different producing countries have reared trees especially for the production of one particular oil.

Thus, the best and most prolific producers of petitgrain oils are the green portions of Paraguayan orange trees. The orange trees in France, Tunisia and Morocco are cultivated for their blossoms, from which come neroli oil and orange flower absolute. Bitter orange oil itself comes from countries which are the best producers of the fruit.

Together with the other Citrus notes, bitter orange oil forms the fragrance prelude for many fresh perfumes. In this role it is not fragrance-determining, but it is indispensable.

Petitgrain oil lies between neroli oil and bitter orange oil price-wise. It is used in Eau-de-Cologne and other compositions that feature freshness as an essential characteristic.

Tuberose: a Fragrance Worth its Weight in Gold

The tuberose or night-hyacinth (French: tubereuse), which resembles the lily and members of the freesia family, is especially grown for the perfume industry. The main growing areas are the south of France as well as Italy and Morocco. The production of tuberose oil is so expensive that every drop is literally worth its weight in gold. One kilogram of the oil is worth approx. 10.000 – 20.000 Marks, and the annual world production is very small.

The blossoms are individually hand-picked, wrapped in damp cloths and immediately proces-

sed, often by *enfleurage*–extraction with lard–a process that is becoming increasingly rare. In *enfleurage,* the blossoms are pressed into the fat, which is smeared on each side of a sheet of glass. They remain in the fat for two days, and are then removed and replaced with new blossoms; and this is repeated until the fat is saturated with fragrant substances. Through a further extraction with alcohol, tuberose pomade oil is obtained.

A more efficient and less expensive method is extraction with a liquid solvent. Even so, the tuberose absolute remains one of the most expensive available to the perfume industry. It is used only in very small amounts, and only in the most prestigious compositions.

Rose Oil: Great Effort for a Great Fragrance

When the Bulgarians speak of something really precious, they'll refer to it as being "as expensive as rose oil." The largest single rose cultivation area in the world is in that country, at the foot of the Balkan Mountains.

The Damascene Rose, a rose similar to the hedge rose, has been cultivated in this area for centuries. It blooms for just thirty days each year. The blossoms are still picked individually, as they were in olden times, and the picking is always a race against the sun. As the temperature rises, the rose blossoms lose their essential oils; and the rate

of loss rises steeply. At noon, for instance, the blooms are only half as rich in oil as they are at dawn. In the high season, whole families–children, grandparents, uncles and aunts–come to the fields to pick the blossoms. They put them in bags that hang down in front, slung from the neck. The really experienced people can pick up to fifty kilograms a day. That may sound a lot, but the yield in rose oil from this quantity of blossoms is only a few drops. It takes approximately five tons of blossoms to produce one kilogram of oil. Each day's harvest must be processed within twenty-four hours. It is distilled to rose oil in government plants and costs around forty-five hundred dollars per kilogram on the world market.

Centifolia roses cultivated in France, in the vicinity of Grasse, in Egypt and in Morocco, have different fragrance qualities from the Damascene variety. In the case of Centifolia blossoms, distillation is not used in the processing, but extraction; and the product is rose absolute. Both rose oil and rose absolute are main contributors to the fragrances of a variety of famous, high quality perfumes.

A Variety of Fragrant Odors – from Wood and Grass, Glands, Moss and Resins

Sandalwood, a Perfumer's Jewel.
Sandalwood possesses unique fragrance characteristics and has a special position in the perfume industry. The oil extracted from the roots and the heartwood of sandalwood trees, particularly the East-Indian Santalum Album, is to be found in every perfume laboratory. Besides its special woody character, it gives a composition warmth, richness and originality. Its effect is insinuating, powdery, rounding-off, fixing. It gives a balmy fullness and, in certain circumstances, animal-like nuances. This wood, which has been popular for thousands of years as a fragrance-source, is so precious that it is gathered and processed under government supervision, mainly in the East-Indian province Mysore (where over 85% of the Indian sandalwood is grown). The trees must grow for thirty years to a height of ten meters approx., before their oil is fully developed. When they are cut down, the wood and roots are pulverized and processed to essential oil by steam distillation in modern plants.

Vetiver – a Woody note derived from Grass.
The woody tone that is used to round off warm, spicy and masculine perfumes in particular often comes from the essential oils of roots. Vetiver is a good example. In countries where it grows wild, this tropical grass is used for thatching huts and mat-weaving. Vetiver oil, which has a strong earthy scent, is distilled from the roots. The oil itself does not have the importance it once enjoyed in the perfume industry, however. Today, it is the most important raw material in the

processing of the so-called vetiverylacetate, which has an elegant and distinctive woody odor.

Musk – the Priceless Warmth.
The legendary musk, one of the oldest of perfume ingredients, is used only to round off the finest of extrait perfumes. The tincture of the sex secretion of the male musk deer, that lives mainly in Tibet, is now beyond price and has been replaced by synthetic products.

Oakmoss holds the Fragrance.
Moss is a lichen which grows on the bark of trees. For its use in the perfume industry, it is scraped off the bark and extracted to give oakmoss absolute or oakmoss resin. These substances are most important fragrance fixatives. They give a composition warmth and extraordinary lasting qualities.

Resinous Balsams for the Foundation.
Resins, such as elemi, opoponax, or Benzoin Siam, from the olibanum or myrrh tree are other natural fragrance fixatives. They come from tropical or subtropical trees and give stability, warmth and richness to the foundation of a perfume.

Jasmine Absolute: 8 Million Blossoms for 1 Kilogram of Fragrance Material

If you want to harvest jasmine, you have to get up early. The blossoms are picked before the dew and the day begin, before they lose their fragrance in the heat of the sun. Jasmine, brought in the 16th Century from Persia to Europe, is still cultivated in the south of France, in the vicinity of the perfume city of Grasse. Here, the rows of plants grow in terraces on the slopes. Harvest-time is from the beginning of August until the end of October. A grower may only have 200 or so bushes – this is about all one family is able to take care of, because the blossoms must be picked every day. By noon, the field is cleared of

blossoms. The next morning, it sparkles again in a dazzling new whiteness. Jasmine blossoms must be processed quickly. They lose their fragrance as their freshness fades.
In the processing, jasmine looks less idyllic than it does in the fields. The blossoms are stacked on strainer grilles in the extraction container. A solvent, such as petroleum ether, is then passed through them. When the solvent has been charged with fragrance, it is distilled. The resulting product is known as jasmine concrete. Through further extraction, with alcohol, it becomes jasmine absolute.

It takes many blossoms to produce a little absolute through this complex procedure. In 1980, the total jasmine harvest of the Grasse region was around 80 tons of blossoms. One kilogram of jasmine blossoms contains approximately 8000 individual blossoms.

One ton of blossoms produces around 2.5 kilograms of concrete, and finally, 2.3 kilograms of concrete yield one kilogram of absolute.

The world consumption of jasmine could not begin to be covered by the production of Grasse

15.M.L

alone. Fortunately, jasmine is also cultivated elsewhere, and on a much larger scale. In Central and Eastern Morocco, for example, on 500 acre-plus plantations using modern methods of cultivation. Here, because of the warmer climate, the bushes grow to over nine feet high. In order to make the harvest easier, they must be cut down to approximately six feet.

While the Grasse jasmine is cultivated and harvested by the traditional family methods, the big, modern plantations operate under industrial conditions with regular employees throughout the year and, during the harvest, with seasonal workers. At these times, three generations of a single family, from Grampa to the youngest child who can walk, might come to work the harvest. In the early afternoon on a typical day, a stream of people dressed in all the colors of the rainbow, comes from all directions and heads for the collection point, laden with blossoms. Here the blossoms are weighed and the wages paid, immediately, in cash. Having received their pay, the workers leave the farm on foot, by truck, or by tractor-trailer. The manager watches them go and hopes–"Inshallah"–that they'll be back again

tomorrow. Whether they will turn up or not is never certain. If there should be a chance of a three-day family feast, they simply won't be back for at least three days. This is the way of the jasmine harvest in all the jasmine-producing countries–such as Algeria, Egypt, France, Italy, Morocco, and, with some differences, also in India, China, and Turkey.

It is not only the Almighty, the stars, and family parties that cause the fluctuations of jasmine crops and ipso facto of the prices. Jasmine is delicate. It can't tolerate frost (which occasionally occurs in the Grasse area), and it intensely dislikes the prolonged droughts that can occur in the African and Mediterranean countries. Further, the blossom won't give up its fragrance substance, the jasmine absolute, unless it is handled with extremely meticulous care. With a precious drop of absolute, the perfumer possesses a gem. Jasmine absolute is an olfactory giant, an essential ingredient of epoch-making fragrance creations such as Arpège/Lanvin, Joy/Patou, Diorella/Dior, Charlie/Revlon, and Miss Dior/Dior. All these perfumes, tremendous successes throughout the world, would have been impossible without jasmine.

Fragrance and Taste: Discoveries and Developments

The fine fragrance of a Muguet scented soap has as little to do with the actual muguet flower as the green color of a citrus soft drink has to do with limes. Neither the fragrance nor the flavor substance is extracted from the real plant or blossom. Instead, they are imitations of Nature called compounds, or "compositions," and they are made from natural, synthetic, or partly synthetic substances. Yet they reproduce the fragrance and flavor effects of their original models so effectively that the user cannot distinguish the difference. Have nose and palate been deceived?

The world of fragrances and flavors would be much the poorer but for the work of the chemists, who began in the middle of the 19th Century to analyze natural fragrances and flavors under scientific conditions, and to isolate their ingredients in chemically pure form. Moreover, they succeeded in producing some of these isolates from different, less expensive starting points. Not content with that, the chemists even created entirely new odorants which do not occur in Nature. A century ago, only 150 fragrance and flavor substances were known; today, more than 4000 have been identified.

Until that period in the middle of the 19th Century, only the ancient methods of distillation, extraction and expression had been known. They had evolved through the previous centuries. These were the methods employed to extract the odoriferous and aromatic elements of plants and animal substances, and to process them further to essential oils, pomades, concretes, tinctures or absolutes.

A milestone on the way to independence from the limitations imposed by the scarcity or difficulty of obtaining natural fragrance substances was the discovery of synthetic vanillin. As a result of a number of research activities that have become classics in the field, Professor Ferdinand Tiemann of Berlin and Dr. Wilhelm Haarmann of Holzminden, also in Germany, succeeded in isolating a glucoside known as coniferin, from the cambial juice of the conifer. Through a chemical process, they extracted from it a crystalline substance identical with vanillin, the main ingredient of natural vanilla, which comes from Vanilla Plantifolia, an exotic orchid.

Haarmann and Tiemann had succeeded in synthesizing the first nature-identical aroma

chemical to be produced on an industrial scale. Only by associating vanilla solely with children's puddings could one fail to appreciate the significance of this discovery. The aromatic principle of the Vanilla Bean–vanillin–is an indispensable ingredient of countless food products that are consumed daily in most parts of the world. Without vanillin, there could be no cocoa, no chocolate, for instance. The products of the cocoa bean are intrinsically bitter in taste. Only the addition of vanillin and sugar produces the "typical" cocoa-chocolate taste we all know.

The first synthetic vanillin was by no means cheap. Only when Professor Tiemann found that it could be produced not only from Coniferin, but also from the very much cheaper Eugenol (a component of clove oil) did the price of vanillin come down drastically. As a result, production and sale increased by leaps and bounds, and chocolate became an affordable luxury.

Vanillin possesses an interesting characteristic: it is one of those substances which has the ability to

PATENT-URKUNDE

№ 576.

GESETZ v. 25. MAI 1877

AUF GRUND DER ANGEHEFTETEN BESCHREIBUNG UND ZEICHNUNG IST DURCH BESCHLUSS DES KAISERLICHEN PATENTAMTES

Dr. W. Haarmann in Holzminden

EIN PATENT ERTHEILT WORDEN.

GEGENSTAND DES PATENTES IST:

Verfahren, das Vanillin künstlich darzustellen.

ANFANG DES PATENTES: 13. Juli 1877.

LÄNGSTE DAUER DES PATENTES: 24. Maerz 1889.

DIE RECHTE UND PFLICHTEN DES PATENT-INHABERS SIND DURCH DAS PATENT-GESETZ VOM 25. MAI 1877 (REICHSGESETZBLATT FÜR 1877 SEITE 501) BESTIMMT.

ZU URKUND DER ERTHEILUNG DES PATENTES IST DIESE AUSFERTIGUNG ERFOLGT.

Berlin, den 20. Maerz 1878.

KAISERLICHES PATENTAMT.

Beglaubigt durch [signature]

Sekretär des Kaiserlichen Patentamtes.

impinge equally on the sense of smell and the sense of taste. Normally, the sense of taste is proficient only at identifying sweet and sour, salty and bitter. It receives other, more specific information through the sense of smell. For instance, the "taste" of freshly-baked bread or fried bacon is perceived more as an odor than an aroma.

This is why fragrance and flavor substances are so often mentioned in the same breath. And why synthetic vanillin is important for another area besides food–the perfume industry. Vanillin is frequently used in the composition of perfume oils. Its sweet, aromatic fragrance is an ingredient of flowery-sweet or powdery, aldehydic, oriental, and woody notes. It is used as a so-called fixative, to hold the lighter, volatile components. Vanillin is also useful in the field of decorative cosmetics, especially for the perfuming of lipsticks; and it is significantly also used in perfumes for baby products.

Coumarin, the principal fragrance- and flavor-ingredient of woodruff, plays a role in the industry similar to that of vanillin. Formerly, natural coumarin was extracted from the tonka bean, the fruit of a butterfly-blossom tree that grows in the northern part of South America. It is so rare and costly that it would be impossible to cover the world's demand from natural sources alone.

In perfumery, coumarin gives nuances to Nature-like "herbaceous" compositions, such as chypre or fougère. Since 1875, coumarin has been produced 100 % synthetically.

Great strides forward were made in the flavor and fragrance industry around the end of the last century. A significant milestone was the synthesis of ionone, the fragrance substance of the violet. It smelled so wonderfully like the real thing that even the elite corps of perfumers, the purveyors to High Society, were forced to abandon their prejudices against the new synthetic fragrance materials. The initial scepticism regarding "Fragrance from the Retort" eventually just melted away–and there soon came in its place a great enthusiasm for the steadily-growing palette of fragrance specialties the industry was producing. For the first time, the scents of the lilac and muguet flowers could be imitated–and used as perfumes. To this day, fragrance experts are still unable to extract their natural counterparts in commercial quantity. Above all, the new odorants which had no counterparts in Nature were fascinating to the perfumers. In the 1920's it was the aldehydes which made the creation of the new, modern perfumes possible, among them Coco Chanel's hugely successful N° 5. This was just the prelude to a profusion of brilliant compositions which owe their diffusive powers and originality to these so-called "Children of Chemistry," the aldehydes. The industrial production of fragrance substances did not change the world, of course, in the way that inventions such as the automobile did. But it could be said that in its own way, the art of fragrance material synthesis has had a considerable effect on our lives.

By making available inexpensive, high-quality substances, it has liberated Perfumery and made this beautiful, enjoyable thing affordable for the broad mass of users. It also freed the perfume industry from its dependence on the declining resources of Nature. The synthetic fragrance industry was honored officially, probably for the first time, in 1900, at the Paris World Fair. On this occasion, vanillin and ionone were singled out for special mention, along with such achievements as aluminium and aniline, in a review of the great scientific and technical discoveries of the preceding century.

Perfume oils, the fragrance compositions of fragrance material producers, go not only to the haute couture perfume houses, but also into many different kinds of consumer-goods such as cosmetics, toiletries, soaps, laundry detergents, cleansers and other household products, and also into rubber, plastic, paper and textiles. Whatever kind of consumer-product we may take in hand, be it furniture polish or a paper handkerchief, the chances are that it has been made more pleasant to use, through the work of the creative perfumery industry.

The nose of the civilized human being has become more fastidious and demanding. All the rose, mint and lavender fields of this world would not be able to fulfil the needs of the hygiene and body-care industries alone. The extraction of natural fragrance substances is difficult and expensive, and the cultivation areas are becoming smaller every day because of soaring labor costs and countrymen being drawn to the towns. On top of this, areas still untouched by population growth, road construction, erosion or pollution problems are needed for food production–above all in countries of the Third World, where most of the exotic essential oil bearing plants are grown.

The mass cultivation which has increased the world's agricultural production so dramatically can also lead to deterioration of taste. Anyone who can still remember the scent and flavor of wild strawberries, field mushrooms or wild marjoram knows it–and feels the lack. Indeed, the cultivated cousins of those strawberries, mushrooms and marjoram plants produce larger harvests, but less flavor. Concentrated flavors, called essences, produced by the flavor and fragrance industry, provide a constantly flavorful taste in a great number of foodstuffs. In some countries only natural and Nature-identical substances are allowed to be used in food products. The "nature-identical" flavor ingredients are synthetic substances that duplicate their natural counterparts completely in terms of chemical structure as well as fragrance and taste characteristics.

A good example is menthol, the main ingredient of the essential oil of mint. Synthetic l-Menthol matches the natural so faithfully that it qualifies worldwide for use in pharmaceuticals. Menthol is one of the most widely-used flavor and fragrance substance that exist. From chewing gum to cigarettes, from cough-drops to toothpastes, from pills to paper handkerchiefs, scores of consumer products are enlivened by aromatic, refreshing, cooling menthol.

Even if there were enough natural fragrance and flavor raw materials to cover the world's needs today, there would be still another problem: many of them would be too expensive to be of practical use.

The price of a kilogram of jasmine absolute goes to five figures. Moreover, Nature has a mind of her own: the quality of a fragrance substance can fluctuate drastically according to the climate and condition of the soil in which it was cultivated. Furthermore, bad harvests, political conflicts, exhaustion of the soil or transportation difficulties are imponderables which make it impossible for the perfumer to rely entirely on Nature's raw materials. Against that background, synthetic fragrance substances appear as economically indispensable substitutes for Nature's originals. They have put the whole fragrance and flavor business on a base so broad as to be unimaginable a few decades ago.

The modern fragrance and flavor industry is capable of meeting three basic requirements of manufacturers producing high-quality branded goods: Supply of any desired quantity, stable prices and–most important, this–a guarantee of constant quality. Except for a few exclusive lines which are of little consequence in terms of the mass market, perfumes today are produced in huge quantities. In spite of this, each tiny bottle must contain the same scent that made it famous years ago. Or, for example, a soap: bar by bar, by the million, each one must have the fragrance that persuaded the consumer to make the initial purchase and the repeat purchases, out of loyalty to the brand.

Even though the fragrance business has its "nose" in countless products that we use every day and every week, it is still one of the least-known of industries. While there are a few makers of perfumed consumer-products who compose their own fragrances, most major manufacturers call on the fragrance industry to produce their perfume oils for them.

It is an essential characteristic of the industry to remain in the background. The perfume "L'image de nuit" or the soap "Morning Fresh" carries its brand name. That's enough. The name of the firm that created and supplied the fragrance is known only to a privileged few.

The quantity of perfume in the finished product is usually minute. For example, one kilogram of perfume oil would be enough to perfume one thousand tablets of soap each weighing 100 grams. The cost of the perfume in each piece is practically of no consequence. Yet the success of the soap "Morning Fresh" would be unthinkable without the fragrance.

One might get the impression that chemistry has outdone Nature, that flowers from the retort might be superior, somehow, to those from the garden. This is true only in a few cases, because most of the natural blossom and plant oils are so complex that it has not been possible to unlock their structures completely. However researchers in the laboratories of the fragrance and flavor industry have brought up some really heavy artillery for their assaults on Nature's secrets: Gas

It used to be that large areas of fertile soil had to be withdrawn from food production for the cultivation of peppermint. Large areas of the important subtropical rain forest were cleared, the trees burned on the spot, and the areas used to grow peppermint. In five years at most, the soil would be exhausted and new areas would have to be found. This depletion of natural wealth has been substantially reduced through the research efforts of the flavor and fragrance industry. Menthol, a constituent of peppermint oil, has been fully synthetically produced since 1973 on a very large scale.

chromatography, mass, infra–red and nuclear resonance spectro-photometry–super-sensitive, sophisticated and expensive devices are part of the analytical arsenal.

Once the particular substances are found and their structures are identified, the next step is the development of suitable (i.e., economical) processes for their synthetic production. But that is only part of the research endeavour. There is a never-ending quest for entirely new fragrance components, not found in Nature, which will enable the perfumers to imbue their creations with original, unprecedented odor-effects.

The fragrance industry still has much to accomplish. The need to develop inexpensive and high-quality substitutes for rare and expensive natural products is more urgent than ever–it is demand and challenge all in one.

"Patchouly," said a perfumer, musing, "if we could only find a way to duplicate that odor, even half-way . . ." Even though the unprepossessing stinging-nettle-like plant with the sought-after fragrance material has given up a large portion of its secrets to the chemists, an economically viable synthesis of its essential ingredients is still not in sight. "When we're able to do this," said a creator of choice fragrances, "we will have found the Philosopher's Stone."

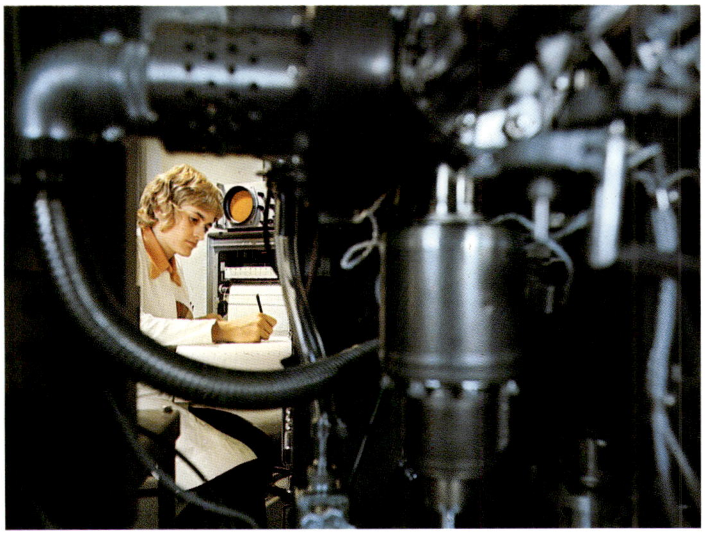

Vision of a Fragrance: from the Original Concept to the Finished Perfume Compound

In the beginning, there is an idea, the vision of a fragrance. It could be the concept of an extrait perfume, or perhaps the fragrance for a whole range of cosmetics and toiletries, from face cream to soap to bubble-bath. Whatever it may be, the idea must be translated, step by step, into the industrial mass production of the perfume oil.

A creative perfumer can see the rough outlines of a new fragrance formula in his imagination before even beginning the actual work of composing. He works with his highly-developed sense of smell as the musician works with his absolute pitch. And like the composer, who every so often plays a few notes on the piano, the perfumer checks on the progress of his composition by holding a smelling-slip, or test blotter, to his nose from time to time. Smelling slips are strips of absorbent paper. They are dipped into fragrance substances, sniffed, and evaluated. By this means one can assess the initial odor of a fragrance composition, observe the changes that occur during the process of evaporation, gauge the tenacity and check what part of the odor-complex finally remains as "dry-out." Smelling-slips are also used in the daily "smelling sessions"–odor-training that is as important to the perfumer as exercises at the bar are to the ballet dancer.

The next step is the formula. "Fragrance Components" is the technical term for the ingredients of a perfume oil–from the uniform, chemically pure fragrance substances to the complex bases, the already-composed basic accords of a fragrance

mixture consisting of natural and synthetic fragrance substances. The finished product usually consists of 30 to 50 such components. But there are also compositions which contain several hundred.

When the perfumer has completed his first trial formula, the ingredients are weighed and mixed in the laboratory precisely according to his handwritten instructions. The perfumer and his assistant draw the ingredients from the so-called Fragrance Organ, tiers of shelves which hold as many as two thousand little bottles of fragrant substances–expensive ones and cheap ones, materials that are rarely employed beside the staples in every-day use–all very democratically arranged in alphabetical order. Some materials deteriorate at room temperature and are therefore stored in refrigerators. Others, which are extremely viscous, are kept pourable in heated cupboards.

Each day the perfumer returns to his composition, smelling and adjusting the formula to meet the stated requirements. Is the composition harmonious, radiantly diffusive, well equilibrated? What is missing? How can it be improved? What do the colleagues think? When the composition is more advanced, we enter a stage of successive

H&R
HEDION

H&R

re-appraisals alternating with formula adjustments: a little more emphasis on ingredient A, tone down the obtrusive B, leave out C altogether, and finally round and rationalize to present a homogeneous fragrance complex.

Next the perfumer must determine whether his composition is compatible with the end product. Whether the composition is for a perfume, a cosmetic range, a soap, laundry powder, hand cream or dishwashing detergent, its behaviour in all the media for which it is destined must be checked. How does the fragrance suit the product? Has either one changed, deteriorated? Is it stable? And, above all, can the perfume be produced at the price the customer wishes to pay? The perfumer must take all these questions into consideration. Like the composer with his instrumental arrangement, he must ask himself whether the work can be performed at all, and, if so, for what kind of (paying) audience. A perfumer must be eternally cost-conscious. To continue the comparison, if he has created a composition consisting only of amber and tuberose, he will undoubtedly fail, as a composer will fail who insists on using only first violins.

Once the finished composition has passed all tests and been approved, bulk production is organized on a scale that is difficult to relate to a small bottle of perfume. The formula is issued to the fragrance company's compounding rooms where ingredients are weighed and mixed by the kilogram or the ton instead of grams, as in the laboratory.

In the midst of all this activity is the computer, which monitors, memorizes and prints out the single ingredients of the composition and indicates the market prices of each one of the fragrance components.

Some of the ingredients of the perfume oil, in kilogram and ton quantities, are drawn from huge tanks, automatically through the use of computers; while others required in smaller quantities are drawn from drums or bottles. And yet, the product which is now despatched by plane, truck, ship or train, is the exact duplicate of the original composition. If the fragrance happens to be of the "fresh as a Spring Morning" type, then a bulk delivery of Spring Freshness is on its way. . .

Perfume: from the Concentrated Composition to International Brand

Perfume, in the finished form which finally arrives in the hands of its purchaser, means not only the fragrance concentrate and the solvents but also–and perhaps as importantly–a beautiful bottle and package.

Prior to packaging, perfume oils and compositions are stored in large bottles and canisters in the store rooms of the perfume manufacturer until they are mixed with alcohol and water, according to their secret recipes, to produce the finished products. The water used for this is especially purified. After mixing, the finished products must mature for periods of up to three months, depending on the particular formula. Perfume is matured in steel containers at a temperature of eight degrees celsius in a dark, temperature controlled room.

It then goes through another of its many filterings before being sent to the filling station. There another quality control check is made, and the filling into bottles takes place. It is done automatically, with the bottles passing by on a conveyor-belt. The stoppers are inserted next. Depending on the particular brand and application, they are either plain or contain tiny tubes for spray appli-

cation. The stoppered bottles are checked once more, this time by an automatic indicator, to insure that they have been correctly filled. Next, the labels are applied, and the final decorative caps. The bottles then move to the packaging machine, where they are put into boxes. Gathered together into packages of six, they are wrapped in cellophane and then moved to the warehouse for shipment and, finally, sale.

In this automated process, the bottles are touched by human hand only for control purposes, or for mounting metered valves or the fragile, ground-in glass stoppers.

In the early days, perfumes were stored in wooden barrels, however, the steel tanks of today, already referred to, do the job without any loss of quality.
Similarly, the last filtering stage, which removes any cloudiness which might remain in the liquid, used to be done by hand, with paper filters. This method is still being used in smaller factories. But in most modern installations, filtering is now fully automated.
It's a very long road, and only at the end of it, when the flacon is opened and the contents are applied to milady's skin, does the perfume come into its own. . .

Sequence of Operations in the Production of Alcoholic Perfumery

Alcohol | Perfume Oil | Distilled Water

Mixing
(At room temperature)

Maturing ▶ (In storage for about 4 weeks)
Cooling ▶ (At 0° C)
Filtering ▶
Filling ▶
Packing

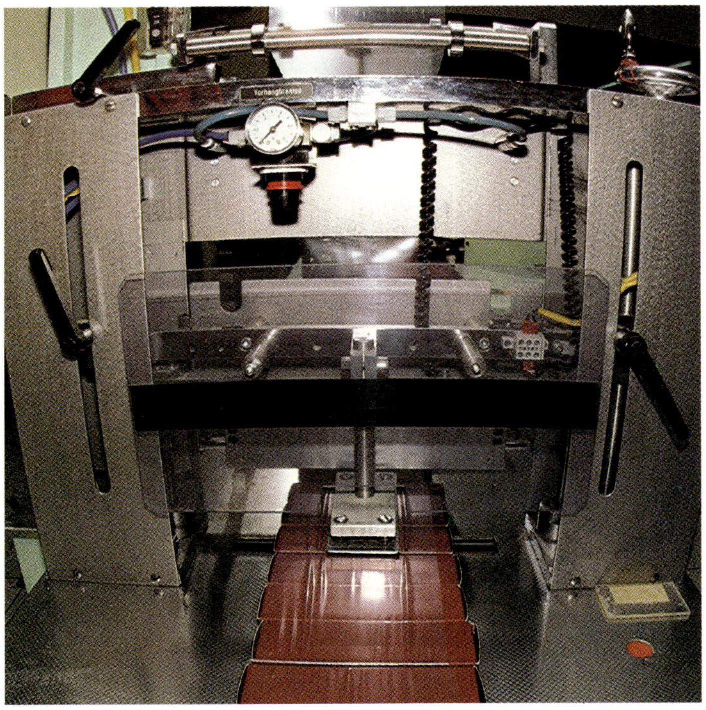

Fragrance from the Psychological Side. The End of a Myth?

Even in times of economic stagnation, the perfume industry keeps growing. At least a part of the reason must be that the people of today have a powerful need to perfume themselves, more and more frequently.

Why?

In view of today's hygienic facilities, the motivation is probably not as it used to be–to mask unpleasant body odors. Apparently, people want to use fragrances and perfumes to satisfy needs that they sense within themselves.

What are these needs, and how are they formed?

The results of recent developmental psychology findings show how they are developed over the course of a human lifetime.

Fragrance and Perfume Needs are Learned.

Personality development from infant to adult can be characterized by the individual's increasing number of needs. Satisfaction of these needs offers contentment, happiness, inner strength and power.

Put into hierarchical form, these needs display a development from physiological needs, such as hunger, thirst, etc., to "psychological needs," such as the need for security, respect, recognition or self-realization.

Before an attempt can be made to achieve the next-higher psychological need, it is first necessary to satisfy the physiological needs.

For an infant, satisfaction of its primary needs is the main objective. The infant's sense of smell is of particular significance in connection with the intake of food. Thus, for example, an infant is able to recognize the odor of its mother's breast from the 6th day on. Since being fed and being cared for are perceived by the infant as being enjoyable, the feeling of enjoyment is also applied to the odors connected with these activities. As a result of this linkage, fragrances and odors become enjoyed sensory stimuli. Siegmund Freud, as well as other representatives of the psycho-analytical school, have described an enjoyment of odors among infants and small children and related this enjoyment to the phases of psycho-sexual development within the scope of satisfaction of libidinous drives.

Observations of the behavior of infants also show something that manifests itself fully only in later stages of life: A general interest in fragrances and odors. An infant is interested not only in fragrances and odors that are connected with feeding and care, but is also interested in smelling various other things such as toys, bedding, feces, etc.

Recent studies have clearly indicated that newborn babies have much greater olfactory sensitivity than had previously been assumed.

Among babies, the odors connected with infant feeding trigger suckling activities and the corresponding respiratory and circulatory changes. Preferences for sweet odors can be identified, which can be explained by the overlapping effect of the senses of taste and smell. Newborn infants display a positive taste response (i.e. manifested by means of continuous suckling) for sweet solutions, while they will stop suckling in the case of salty, bitter or acidic solutions and will display a pronounced aversion and dislike.

However, unlike taste, there are no inborn perceptions of bad or unpleasant odors. Only substances that overly irritate the infant's nasal mucous membrane are perceived as being unpleasant or painful.

Two points should be noted: First of all, humans are not born with a natural sense of bad or unpleasant odors; the exact opposite is true. What is in later life assessed as being an unpleasant odor is the result of the learning experiences of socialization, and not hereditary odor criteria. Secondly, human beings display a general interest in fragrances and odors right from the beginning. They have an inborn need to smell everything possible. In most societies, this need is channelled by the educational process and is encouraged in only few areas; usually, it is suppressed or downgraded as being insignificant.

The dominating needs of *the small child* are security, protection, love and tenderness. At this stage of development, fragrance and odor become an expression of human warmth and security. In order to assure itself of the affection of its parent figure, and thus to comply with its parents' desire for hygiene and cleanliness, the child attempts to observe the rules of cleanliness. In doing so, it learns what smells good or bad. At the same time, the child also develops its own value standards for good or bad odors. This is done by associating given fragrances and odors with pleasant or unpleasant persons, situations and experiences.

With the child's growing need to associate with its peers and with older children, it also learns society's odor standards.

Through identification with an adult model, the child's interest in perfumes, etc. can be awakened at even a very early age. Along with the desire to emulate an aunt, for example, the child will imitate her cultural and esthetic standards. On the other hand, through identification with a person who dislikes perfume, a correspondingly negative attitude toward the field of fragrances and perfumes can also be instilled in a child at a very early age.

With adolescence, the individual attempts more and more to view himself or herself as a distinct personality, and to define himself or herself in comparison with others. In the adolescent, the need grows for respect and recognition, especially among his or her coevals. The adolescent increasingly begins to discover his or her personality, its traits, moods and feelings. Through the adolescent's own observations and through the comments made by others, he or she also perceives bodily sexual development. This includes the development of an individual, sexual-typical body odor. In olfactory terms, the sexes cannot be determined until puberty. In the case of infants and small children, it is not possible to differentiate between boys and girls on the basis of odor. At puberty, the apocrine sweat glands begin to function; they produce a secretion that is very rich in protein and that, when decomposed by microorganisms, determines the individual's personal odor.

For the adolescent, the area of fragrant materials and perfumes is now enriched through new aspects. Fragrances and odors gain in significance for hygienic reasons, to mask unpleasant human odors. Moreover, through perfuming, the adolescent can underscore his or her sexual-specific odor, and thus the usually long-awaited sex characteristics. For an adolescent, perfuming himself or herself represents the feeling of being an adult, a man or a woman, and having left childhood behind. For the same motives, smoking and the use of perfume often go hand in hand during this stage of development.

But for the adolescent, wearing perfume not only means being an adult, but also represents a way of underscoring individuality. Especially in our modern, anonymous society, the adolescent feels forced to call attention to him- or-herself, in order to avoid being lost in the crowd. On a more or less conscious plane, the adolescent who is interested in fragrances learns to employ perfume in order to call attention to his or her personality and gain respect. In selecting a fragrance the adolescent cares less about what suits personality, moods and feelings, and more about which products will produce recognition, especially on the part of the adolescent's peers. This could be the newest product, or a totally unusual fragrance, or the most expensive fragrance, one that others cannot afford.

As we grow older, the needs for self-realization and self-actualization become stronger and stronger. With these two terms, we mean a tendency to continue development in two directions. First, with respect to the individual's own relationship to others, to the environment and to society (self-realization), and second, with respect to the relationship to him- or-herself (self-actualization). In the latter case, self-actualization is the desire to experience and to identify personality, feelings, moods and interests, in order to determine for one's self the development of the ego. In the former case, self-realization means introducing one's self into the environment as a whole person, without giving up or losing individuality. Moreover, weighing all of the interests, to find a place in society for one's self that corresponds to the specific inclinations of the individual personality.

In accordance with these two directions, perfume serves two purposes for the individual at this stage of development. Perfumes offer the possibility of being able to identify and underscore the individual's moods and feelings. They thus activate the way the individual's own personality is experienced. In the same manner, the individual can use perfumes to create moods and feelings that he or she seeks or desires. Thus, perfumes have stimulating and corrective effects on moods. As a possibility for self-actualization, they become part and parcel of the individual's own personality.

At this stage of development, perfuming one's self means satisfying the fragrance need that results from what is being experienced at the moment. Moreover, the fragrance and perfume area is the subject of major significance in conjunction with self-actualization. Perfumes become an expression of individual style, in which they underscore the external manifestations of the individual's personality. Perfumes thus define the relationship between the wearer and the environment. They are an indication of the individual's cultural and aesthetic perception. Perfumes become a message that necessitates a certain degree of sensitivity on the part of the receiver and a willingness to relate to the sender if the message is to be decoded. Thus, as a means of non-verbal communication, perfumes can awaken sensuality and empathy in others. However the decoding process can be equally likely to contain an invitation for activity, vibrance and joy.

The need for self-realization and self-actualization now also manifests itself in a change of attitude toward perfuming. At this stage, a fragrance aesthete selects those perfumes which, through their radiance, best underscore his personality in the context of the specific situations.

In asking ourselves what the meaning of perfume is for people today from a psychological standpoint, we must always bear in mind the individual's odor and fragrance socialization process (the learning experiences made during the process of growing up).

Every individual develops a distinct relationship to the area of odors and fragrances in each developmental stage.

These odor and fragrance experiences, which are linked to feelings, emotions and emotionally colored experiences, are stored in the limbic system via a personality-specific stimulus processing step. Individual odor and fragrance preferences can be explained by the stored fragrance and odor experiences and the corresponding associations gained over the course of the individual's personality development.

The odor and fragrance experiences of childhood are very critical with respect to the adult's relationship to the area of fragrances. Thus, upbringing processes can impair, or entirely suppress, odor and fragrance experiences made during the individual development stages. Consequently, the adult striving for self-realization and self-actualization cannot employ perfumes as media, as he or she was not able to experience the fascination of perfumes during the developmental years. The widespread opinion that this inability is caused by an inborn or learned reduced olfactory power is, when subjected to closer investigation, not very convincing. Like hardly any other sensory organ, the sense of smell can be trained.

The possibility that reduced olfactory perception is caused by physiological and pathological circumstances is highly remote. Moreover, recent medical research work has even identified regeneration processes in damaged olfactory cells.

Independently of their age, adults can be classified into one of the following four stages on the basis of their individual fragrance and perfume learning experiences. Each of these stages results from characteristic attitudes towards odors and fragrances.

1. The soap users.
This user group claims to have no particular connection with fragrances. For this group, the fragrance area plays a role virtually only in conjunction with eating and drinking. All other fragrance areas are only of secondary importance. Soap is used almost exclusively for personal hygiene. People with this attitude do not perfume themselves.

2. The fragrance gift users.
Fragrances and perfumes are not used to satisfy a personal need, but because their use is desired by the individual's partner, for example. These individuals usually receive fragrance in the form of gifts, or such products are bought for them. Fragrances give them the feeling of not seeming unpleasant to surrounding noses. Nevertheless, fragrance gift users usually perfume themselves for specific events, because others do it, too, or because it is customary to do so.

3. The brand-name perfume users.
For these individuals, high standards of hygiene demand brand-name fragrance. Perfumes signal the individuality of these types to themselves and to others. For this group, wearing perfume means being able to afford more luxury than others. In addition, it is also important to them that perfumes exert an attraction on the opposite sex and awaken interest in them.

The members of this group perfume themselves in order to define themselves in comparison to others and to demonstrate who they are, how they would like to be, and what they can afford.

4. The perfume aesthetes.
For perfume aesthetes, fragrance creations are media for self-actualization and self-realization. They select those perfumes that correspond to their personality in terms of what they are experiencing at the moment – their feelings and moods.

At this high stage of fragrance experience, perfuming one's self means satisfying one's momentary fragrance and perfume need.

Our Fragrance Memory.

New findings from the fields of sensory physiology and emotional psychology indicate the significance of moods and feelings for fragrance perception and appreciation.

In humans, the olfactory cells are located in the approx. 6 cm^2 (3/4 sq. in.) large regio olfactoria in the roof of the nasal cavity. When the receptors are stimulated, the stimulation is advanced directly to the bulbus olfactorius via bundles of axons (nerve fibre). The bulbus olfactorius is a major structure in the limbic system. Only a few decades ago, the functional significance of the limbic system was completely unknown. Today, we know that this system serves as the functional circuit for emotions, moods and feelings of motivation and sexual behavior, for example, and can be correspondingly stimulated via the sense of smell. Moreover, the limbic system also plays a significant role in selecting and transmitting information between our short- and long-term memories. Selection of and transmission to these two memories is performed via corresponding associated regions of the limbic system that receive their information from the various sensory regions such as the sense of smell. It is thus understandable why perception of certain fragrances can be connected with a recollection of specific experiences, or why, when we recall childhood odors for example, we simultaneously perceive our past emotions. The olfactory receptors are so closely linked with the limbic system that a malformation of the limbic system can frequently manifest itself in the form of faulty corresponding sensations in the olfactory system as well. Thus, the effects of schizophrenia for example, very often manifest themselves in the form of olfactory hallucinations. The patient frequently smells strange odors and suspects toxic gases, which can greatly increase his or her fears in conjunction with paranoia.

In other psychic disorders, such as depression, for example, the interaction between emotion and fragrances lends itself very well to therapeutic utilization. Since time immemorial, first "medicine men," then lay medical practitioners, and today, physicians and psychologists, have attempted to intervene therapeutically in disorders of this nature through the use of fragrances.

The correctness of this approach is also demonstrated by recent research in the clinical-psychological field. Even though it has thus far not been possible to causally treat these kinds of illnesses in this manner, patients always display a much higher assessment of their own well-being after odorants have been administered than control patients who are not treated with odorants. This is a major factor in the healing process, and in a hospital environment in particular, the loss of the subjective feeling of illness is a major step toward recovery. Generally speaking, an odorant therapy is indicated if activity-enhancing, more self-secure and more optimistic impulses are to be aroused in patients in order to improve their general physical well-being, their relaxation, emotional stability and performance and concentration abilities. Thus, fragrance therapy is an ideal accompaniment for psychotherapy.

Are Fragrances selected by Chance? The End of a Myth.

In the past, the significance of psychological and physiological factors in the selection of fragrances was given varying degrees of attention.

Representatives of the "classical approach" felt that selection of fragrances was made on an irrational basis, and could not be explained on the basis of specific factors–i.e., it was held to be a purely random occurrence and thus not psychologically definable. Another opinion, which is still adhered to by many today, cites external characteristics, such as hair and eye color, as being the decisive factors.

In recent years, an attempt has been made, primarily by chemists and perfumers, to use fragrance descriptions in order to make statements regarding perfumes and their users. In doing so, it was found that there was a lack of information about various personality types. In addition to psycho-physiology, psychological aesthetics research also supplied information on personality types. One of the subjects studied by this field of research includes an investigation of preferences for given "aesthetic stimuli" with respect to their relationship to individual psychic traits.

The aesthetic needs and the personality of an individual are very closely related. This manifests itself, for example, in the form of home and furnishing preferences (choice of color, shape and style), in the form of music and art preferences, and so on. From the standpoint of psychological aesthetics research, a fragrance creation–as an aesthetic stimulus–enjoys a very particular status. As mentioned before, the olfactory receptors are directly linked with the emotional centers in the limbic system. Thus, the olfactory stimulation is primarily of an emotional nature. Only on a secondary plane are the cognitive-intellectual aspects of a person (self-perception, attitudes, associations, etc.) affected. But in researching and explaining individual aesthetic preferences in the area of fragrances, this means that particular significance must be placed on personality differences in the emotional area.

At the moment, differentiation between the emotional and the cognitive-intellectual aspects of a person might undoubtedly appear to be somewhat strange. However this differentiation is easy to understand if one's rational intellect (head) is diametrically opposed to one's perceived emotions (stomach). Or if the one or the other aspect is affected more strongly during aesthetic enjoyment. In the research process, clarification of both aspects necessitates different kinds of instruments.

Empirical studies that investigate specific fragrance preferences of perfume users with respect to common psychic features, have produced the following results: There is a close relationship between certain feelings and moods and fragrance preferences.

In explaining fragrance preferences, two personality dimensions are of significance: emotionally stable versus emotionally ambivalent, and emotionally oriented toward the outside, toward the environment (extroverted) versus emotionally oriented toward one's self (introverted).

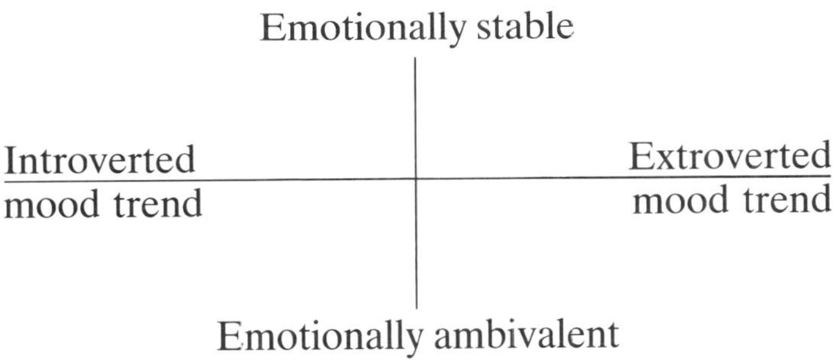

The behavior of the extremes of these two personality dimensions can be described as follows:

The typical *extroverted personality, emotionally* oriented toward the outside, toward the environment, is dynamic and sociable. This personality likes functions, has many friends, needs people with whom to speak, and prefers not to be alone. It wants suggestions, and also likes to give suggestions, always takes advantage of favorable opportunities, frequently is spontaneous, takes risks and is generally impulsive. This personality likes to laugh, always has a quick answer on the tip of its tongue, and loves change. This is a carefree personality, optimistic, cheerful, and likes to talk about itself. This personality likes to move, loves activity, and sometimes, can be almost euphoric.

The typical *introverted personality, emotionally* oriented towards itself, is quiet and tends to be reserved and withdrawn. This personality keeps its distance, except among very close friends. It tends to make plans in advance, is cautious, careful, and mistrustful of spontaneous impulses. This is a personality that does not like excitement, takes daily living with a goodly portion of seriousness, and values a well ordered lifestyle. It considers itself to be more sensitive than most people, and relates more strongly to its own inner world, in which it likes to dwell.

The typical *emotionally ambivalent personality* experiences and shows itself in highly differing moods. And this personality's moods and feelings can change from one moment to the next, which need not necessarily be experienced and perceived in a negative manner. To the environment, this personality often appears dreamy, and incapable of withstanding psychic stresses, sometimes even melancholic. This personality likes to think about things, even if they cannot be realized. It has an aversion to everything that is very sober and rational. It does not want its moods and feelings to be restricted. Thus, in certain situations, it can intentionally feel differently than the environment expects of it.

The typical *emotionally stable personality* experiences and shows itself in well balanced moods and feelings. It attempts to emotionally "get a grip on" the highs and lows in its life. This personality does not like to see itself in an unhappy or depressed mood. It attempts to appear harmonious and well oriented to its environment. In doing so, this personality offers a realistic, purposeful and emotionally strong impression. It can adapt its moods and feelings well to momentary needs.

Individual differences in these two personality dimensions are caused, according to researchers who follow the school of the English psychologist Hans-Jürgen Eysenck, by inborn differences in neuronal activation processes in the limbic system and an adjacent region of the brain. Stated in simpler terms, these are processes of the central nervous system that are determined to a significant extent, and characterized, through an individually differing relationship, by nervous excitation and inhibition processes.

Neuro-physiological processes are thus a decisive factor in the way the respective individual personalities appear in the areas of extroversion/introversion and emotionally stable/emotionally ambivalent. These neuro-physiological processes are responsible for the fact that extroverts generally prefer a higher degree of stimulation than introverts, and need this stimulation for their well-being. Thus, extroverts have a preference for stimulating color combinations, such as red and orange, while introverts tend toward subdued combinations, such as various shades of violet. Conversely, there is a correlation between the fact that extroverts can tolerate higher pain stimuli than introverts. Correspondingly, these processes also cause the limbic system of emotionally ambivalent personalities to respond at a lower stimulus intensity under otherwise identical conditions, while emotionally stable personalities necessitate greater stimulation. Serving, as it does, as the central circuit, monitoring and control point of these neuronal processes, the limbic system, originally also termed the olfactory brain, is of decisive significance. It is responsible for the personality-specific processing of stimuli. Its significance with respect to moods and feelings has already been discussed, as well as its odor and fragrance memory. We also know that the nerves of the olfactory receptors terminate directly in the limbic system, which explains why it was originally called the olfactory brain.

All of this psycho-physiological evidence makes it impossible to continue to believe in the myth of "irrational or random" selection of fragrances.

What makes a Perfume Fascinate?
What causes a Perfume to be Rejected?

When people are fascinated by a perfume, this perfume appeals to three areas of their personality. The better a perfume can please these three areas, the more likely it is to be selected.

First area:
Emotions, attitudes and associations.

The emotions of an individual are a very significant factor. They determine the individual's fragrance need. Emotional feelings and moods are indicative of the fragrance preferences.

To a greater or lesser extent, every human being is subject to fluctuations in feelings and moods. This also explains why a person can tend to like different types of fragrances. Thus, an individual can, at times, be oriented toward the outside, toward others (extroverted), even though the individual's basic mood primarily tends to be introverted.

To test the influence of feelings and moods on fragrance need, female perfume users were subjected to a color rosette test developed by the authors.

Seeing colors imparts certain moods and feelings. Correspondingly, certain color preferences are an expression of emotional feelings and moods. This knowledge is based on the classical psychological color test: The color rosette test represents a further development of this classical color test. It was developed not so much for clinical-psychological use, but rather to identify aesthetic preferences that are determined by moods and feelings.

The employment of the color rosette test in the area of fragrances to investigate fragrance preferences and emotional personality traits among perfume users produced the following results:

- Extroverted perfume users have a fragrance need for fresh-floral notes.
- Introverted perfume users have a fragrance need for Oriental notes.
- Emotionally ambivalent perfume users have a fragrance need for floral-aldehyde, powdery notes.
- Emotionally ambivalent, extroverted perfume users have a fragrance need for floral-fruity notes.
- Emotionally ambivalent, introverted perfume users have a fragrance need for Oriental-floral notes.
- Emotionally stable, extroverted perfume users have a fragrance need for chypre notes.
- Emotionally stable, introverted perfume users have a fragrance need for aldehydic-floral, balsamic notes.

Purely theoretically, one could imagine another group of perfume-users, completely emotionally stable, with a specific fragrance need. The research showed, however, that the purely emotionally stable attitude, aspired to by so many people, is impossible to achieve as a permanent condition. The range of environmental influences and personal experiences which have an effect on the psychic disposition of the individual, moving him or her in one direction or another according to mood and emotion, is multi-faceted. A purely emotionally stable condition would require, for example, the seclusion of a cloister or the atmosphere of a meditation center.

Through an understanding of emotional feelings and moods, the fragrance need, and thus the attractiveness of perfumes having the corresponding fragrance notes, can be explained, and, conversely, can also be identified. Moreover, taking into consideration attitudes and associations also provides an impression of the way an individual relates to her fragrance and perfume need, and which associations she has with the individual perfumes in this area.

Second area:
The desired lifestyle.

For users of top-quality perfumes and perfume aesthetes, perfumes and similar products are symbols of individuality and an expression of "being able to afford," or a possibility for self-actualization. Thus, perfumes become an integral part of an actual or desired lifestyle. This area also includes interests of a professional, vocational and cultural nature, leisure interests, travel wishes, style of furnishing, satisfaction with one's life and one's self.

The studies performed on female perfume users showed the following phenomenon: If perfumes are assessed by the perfume users as being fascinating, their fragrance note not only covers the individual's fragrance need, as determined by her emotional feelings and moods, but also correlates, in its perceived radiance, to the user's desired lifestyle. Perfume aesthetes and users of fine perfumes link their emotional feelings and moods to a certain lifestyle which they consider to be desirable, and which they also seek in their perfumes. In doing so, their attention is called to certain products through appropriate advertising messages. However, regardless of the fragrance note, only those perfumes are assessed as being fascinating in which the associated lifestyle is in coincidence with the emotional feelings and moods that are underscored by the perfumes.

Should there be a contradiction, enthusiasm is clearly limited.

Third area:
Situations, occasions, socio-biographical data.

Emotional feelings and moods are subject to certain influences such as situations and occasions; or have a typical characterization, e.g., through age and sex features. Emotional feelings and moods also influence the way situations are experienced. Thus, an extroverted individual acts differently in the office than she does in her leisure time. So does an introvert. Both experience the same situation, the office routine or their leisure time, in different ways.

It is amazing how dependably perfume users take situations, occasions and seasons into consideration in selecting their perfumes once they have identified their emotional feelings and moods, i.e., their fragrance need. On the other hand, this is not surprising, as each and every user knows through innumerable experiences, what she is seeking in certain situations, how she reacts, and the impression she would like to make on others.

If they not only satisfy fragrance need and lifestyle but also are commensurate with the occasion, perfumes become a second skin.

The following example, selected from the study, shows how the reason for a perfume's fascination can be understood by means of a psychological analysis of these three areas.

At the time of the study, Mrs. S was 39 years old and lived in a small city in southern Germany. Perfume "P" (Genealogy of Perfumes 1975/9) had been an "indispensable part" of her for several years. Analysis of her emotional feelings and mood trends by means of the color rosette test showed that she had an extroverted, primarily emotionally stable personality. Correspondingly, Mrs. S preferred perfumes from the chypre area.

Mrs. S characterized herself as being a frank, warmhearted and well-balanced person. When asked about the associations and attitudes linked with her perfume, she described its fragrance note as "harmoniously well rounded and unobtrusive, yet nevertheless interesting and exciting."

Mrs. S works in the family business, being responsible primarily for dealing with customers. Time and time again, she emphasized that her family did not suffer because of the business. When the children were small, she said, she worked less.

Because of her job, Mrs. S pays a great deal of attention to her wardrobe. Although she does not aim to be dressed in the very latest fashion, she feels her clothes should be smart and stylish. Mrs. S drives a powerful, upper middle-class car. "It's fun for me to drive fast now and then, if conditions permit," she says. If Mrs. S has any time to spare, after taking care of family and business, she is busy running a club. "I have to keep everything going", she says, "otherwise nothing gets done".

Mrs. S likes to go on vacation with her family to southern France or the Atlantic coast of France. All in all, she has a positive attitude toward France, especially toward Paris. This attitude also manifests itself in her assessment of French cooking and French perfumes: "I like French cooking very much . . . The best perfumes come from France."

When asked whether it was merely by chance that she wears a French perfume, Mrs. S smiled and said: "I don't think so!"

Asked about the occasions and situations in which she primarily wears Perfume "P," Mrs. S responded: "Actually, I always use it, especially when I'm at work . . . every day . . . It's unobtrusive and yet it has a special radiance . . . a perfume that says 'Hey, here I come' doesn't suit me . . . nor do I want to wear one of those perfumes that everyone has."

The information obtained from the analysis of the three areas shows clearly how Mrs. S is reflected in her perfume descriptions, both as an entire person, as well as in each individual area. She outlines her perfume in virtually the same manner in which she describes, or circumscribes, herself, her needs and interests. Her positive assessment of herself, which she does not state openly, also manifests itself in the way in which she views her perfume. She characterizes it as being interesting, exciting and having a special radiance. Her demand for individuality is expressed by her aversion to well-known perfumes.

Observations made in perfume consulting situations show that perfume users are generally quite satisfied with a perfume if, in addition to the first area, the perfume also satisfies their demands in one of the other two areas. Thus, if the perfume note coincides with their emotional feelings and moods, perfume users are willing to be satisfied with less in the other areas. Ability to meet with the fragrance need is therefore the basic prerequisite for fascination. If a perfume is not able to satisfy this basic need, it will, at best, be perceived as being interesting. And this holds true especially if a perfume does not appeal to the first area, but does appeal to the other two areas.

Perfumes that do not appeal to the user at all in any of the three areas are assessed especially negatively. And if negative associations and experiences, which can extend all the way back to childhood, are linked with this fragrance type, the aversion can increase all the way to repulsion.

New Possibilities for Perfume Consulting.

Through the information obtained in the three areas, it will be possible, in the future, for a beauty consultant to offer a customer a dependable decision-making aid to help her select a perfume. There is no question that most consumers have long desired an orientation aid. Every beauty consultant is only too familiar with the question asked time and time again by customers:

"Which perfume suits me?" and the ever growing number of perfumes on the market makes the situation even more difficult. In the final analysis, all of those imaginative, beautiful fragrance descriptions merely serve to confuse the issue, as they do not say enough about the personality type. While there are perfume user typologies, they do not offer a sufficient degree of exactness and differentiation, describing the personality, merely as sportive, romantic, modern, and so on. In the past, these perfume user typologies have not contributed much toward helping the perfume user identify her emotional feelings and moods, and thus her fragrance need.

Seven perfume user groups.

In analyzing perfumes and their users in all three areas, it was found that there is a total of seven different perfume user groups.

In order to classify perfume users into the individual groups, to identify their fragrance needs, they are requested to select from the color rosette test the rosette or rosettes they like best. They can chose from seven of the eight rosettes in the color rosette test. The eighth, yellow-green-brown color rosette is of no significance for perfume users. It provides information primarily in other aesthetic areas, such as the home and furnishing style. Should a perfume user select more than one rosette, she is then requested to choose the rosette she feels to be the most aesthetic, the rosette that appeals to her most at the moment. Additional important information can be obtained on the basis of the type of rosette(s) selected and the position of the rosettes relative to one another. For example, which emotional personality impulses does the perfume user feel in herself? In addition, what variants result in her fragrance need? If the user tends to like two color rosettes, not only are the perfumes in the respective fragrance groups of interest to her, but especially, perfume creations that unite both types of fragrances. Whether the one or the other type of fragrance dominates can be seen by the color rosette that is finally preferred most, and that is thus indicative of her fragrance need.

Through this type of analysis, the fragrance need of an individual can be identified on a very finely tuned basis. It permits the perfume user to find perfume creations that, as a result of their complex radiance, are not entirely typical of one type of fragrance. In this connection, it is also understandable why a perfume can appeal to more than just one group of perfume users, e.g. Perfume 1979/7 (Geneaology of Perfumes) appeals not only to the emotionally ambivalent group of perfume users, but also to the group consisting of emotionally stable users with an introverted mood trend. However each group weights the radiance of the same perfume differently. Consequently, the user of the perfume has an affinity for both color rosettes. It is possible even to further define a person's fragrance need by asking perfume users the opposite question i.e. to which rosette do they have an aversion. This approach is advisable if the perfume user does not feel that any of the rosettes in the color rosette test are especially attractive. This type of "trial and error diagnosis" avoids putting a customer seeking perfume advice in a situation in which she is confronted by a fragrance type she rejects, a factor that is of considerable importance in view of the

The 7 feminine perfume user groups, personalities, color and fragrance needs

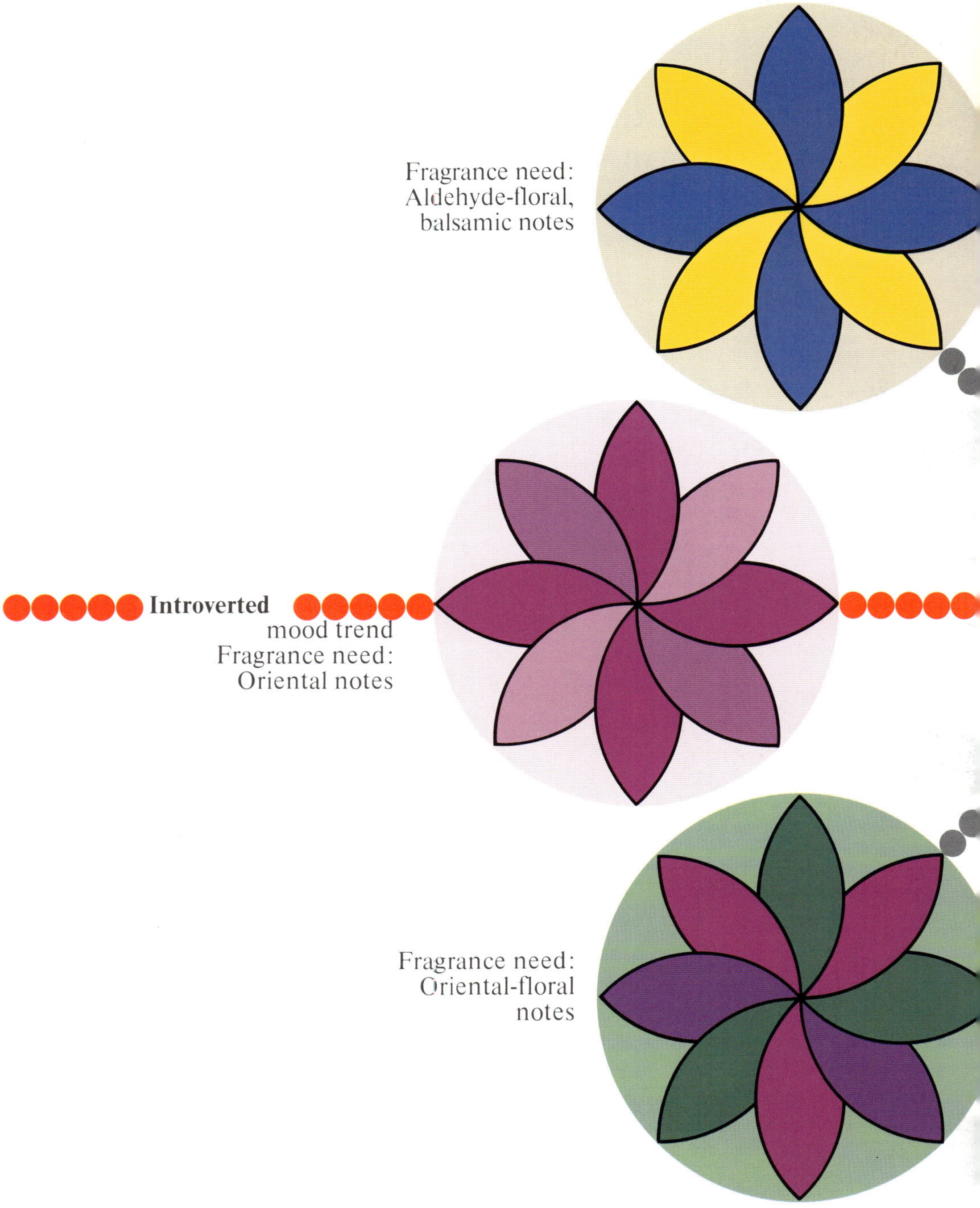

Emotionally
stable
Fragrance need:
Chyypre notes
Group F
Group A
Group D
Group C
Group E
Group G
Extroverted
mood trend
Fragrance need:
Fresh-floral
notes
Fragrance need:
Floral-fruity
notes
Emotionally
ambivalent
Fragrance need:
Floral-aldehyde,
powder notes

fact that the ability of the average perfume user's nose to accept new fragrances usually ceases after she has smelled three or four fragrances. Moreover, the customer feels that her wishes are being understood more quickly, which makes perfume consulting less complicated and time-consuming.

The individual perfume user groups, their color and fragrance needs, and their typical personality traits.

Group A:
Extroverted.

Preferred color rosette.

Fragrance need: Fresh-floral notes.

These perfume users like to live an active life with and in their environment. They are attracted by new things. Their attitudes are optimistic and affirmative. They have overcome difficult phases in their lives through courage and innovation. This group of users would like to work in a field that offers a great deal of variation, the ability to work on their own, as well as travel. And they are willing to accept risks in order to achieve this end. They are in charge of their destinies. The perfume users in this group could never be only housewives. They need success in their work; the word career does not have negative implications for them. They reject the weak, dependent woman's role.

Their desired lifestyle is an expression of the demands they make on life. These women, for example, furnish their homes in a straight-forward, practical, functional manner. They need their homes, to "switch off", enjoy a change of pace, rest and recuperate. And in doing so, anything that stands in their way is an obstacle.

These perfume users are more or less evenly represented in all age groups.

Group B:
Introverted.

Preferred color rosette.

Fragrance need: Oriental notes.

The perfume users in this group want to relate more strongly and deeply to themselves and their environment than most of the others around them. They reject superficial people. These women have no trouble being alone with themselves. They long for peace and tranquility, as well as the possibility of developing their own individuality. They are not very dependent upon the judgement of those around them, but tend to seek their own way. They cannot be forced to do anything that does not correspond to their own feelings. If they are hurt, they withdraw into their own shells. However they can also surprise those around them by representing their interests with self-assurance. This is especially the case if conformity is expected of them and they feel that their individuality is threatened. In this situation, they are very direct in telling those around them what they do not like, or they come back with a ready answer.

Their desired lifestyle is highly individualistic and unique. Clothing, their home and their furnishings, are expressions of an "alternative" lifestyle, as well as their desired self-relationship. They reject anything that appears to be "middle class" or "straight."

These perfume users are generally younger women, or women who feel much younger than their own age group.

Group C:
Emotionally ambivalent.

Preferred color rosette.

Fragrance need: Floral-aldehyde, powdery notes.

These perfume users reject people who view everything in sober, rational terms. They view themselves as being more sensitive and having more finely developed feelings than most of the other people around them. These women frequently think about the possibility of losing their loved ones. They do not like to have their moods and feelings restricted. In social situations, they tend to be somewhat reserved. In no circumstances do they want to force themselves upon others. They also dream about things they cannot realize. These women do not want to view reality in the utilitarian, realistic manner in which it often appears to others. Their imagination aids them in seeing and experiencing things differently every now and then.

These perfume users have great difficulty in parting with old things, as can also be seen by their desired lifestyle. Correspondingly, their home and furnishing style is characterized by items they have grown to love. They are keenly interested in fashion, which affords them security and enhances their self-confidence. Moreover, they also enjoy dressing up in costumes.

In this group, younger women predominate.

Group D:
Emotionally ambivalent with extroverted mood trend.

Preferred color rosette.

Fragrance need: Floral-fruity notes.

The perfume users in this group love life in all of its facets. They enjoy surrounding themselves with excitement and activity, yet also do well when alone. They always attempt to make the best out of what life has to offer.

They demand optimism and affirmativeness of themselves, even if they are not in the mood for it. These women sense that they can be easily hurt, a fact they like to conceal from others. Should their mood or feelings not be particularly good, should they feel injured, they attempt to overcome the situation without brooding over the cause of it longer than is absolutely necessary. They reject people who let themselves go, who are colorless, or who talk only about their illnesses, etc. To others, they sometimes seem dreamy and unaware of the seriousness of life. These women like to indulge themselves; their attitude is: "We should satisfy as many wishes as possible–after all, we only live once." They abhor everything petty. In the final analysis, their feelings play the strongest role in solving problems.

Their desired lifestyle is characterized by spontaneous interests and impulsiveness. They can decide from one moment to another, for example, to go on a trip, and can then promptly change their destination. These women like change and are receptive to new things. These feelings show themselves especially in the style of their home and furnishings.

Group E:
Emotionally ambivalent with introverted mood trend.

Preferred color rosette.

Fragrance need: Oriental-floral notes.

Perfume users in this group reject a hectic, unsettled or dynamic life. They feel best if their lives continue along orderly, quiet, paths. They like to receive suggestions from others. These women enjoy having people around them who can offer them protection and counsel, but without disturbing the rhythm of their lives. Wherever possible, they avoid problems and conflicts. They let others handle things that threaten to upset them too much. Anything that is too loud and garish hurts them. They also avoid day-to-day routine work.

Their desired lifestyle necessitates a firm financial background, which they take for granted. These perfume users like to collect small "gems." In their furnishings, comfort and cosiness rank first. They do not have exaggerated athletic ambitions. While they would not term themselves as being unathletic they do not want sports to become an effort for them.

All age groups are uniformly represented in this perfume user group.

Group F:
Emotionally stable with extroverted mood trend.

Preferred color rosette.

Fragrance need: Chypre notes.

These perfume users view themselves as being harmonious, well-balanced individuals, who rarely have unhappy or depressive moods. All in all, they are satisfied with themselves and with their lives. They are realistic in their thinking and put less faith in their luck and more in their own deeds. When problems arise, these women do not attempt to avoid them, but solve them in a practical, uncomplicated manner, and display self-confidence in doing so. These perfume users are keenly interested in the people around them, with emphasis centering on their private lives. They radiate human warmth and strength to their environment, and give the impression that they are hardly ever in a depressed mood. The users in this group enjoy providing practical help to others and are not infrequently very active in their community.

In their desired lifestyle, these women do not have any interest in keeping up with fashion, in all of its varied manifestations. They love to look well-groomed, and their clothing must always be suited to the occasion. In all matters, especially in the case of clothing, they are highly quality-conscious.

All in all, there are not many very young women in this group of perfume users.

Group G:
Emotionally stable with introverted mood trend.

Preferred color rosette.

Fragrance need: Aldehydic-floral, balsamic.

The perfume users in this group consider themselves to be neither very dynamic and lively, nor predominantly quiet and introverted. Nevertheless, these women do not always live well-balanced lives. Thus, they sometimes have difficulty controlling their moods. And yet, it is mostly they, themselves, who insist that they must not "let go". They feel within themselves that they are also capable of other moods. Correspondingly, it is primarily this group that also has other fragrance needs, as expressed, for example, through the black-and-white rosette in the color rosette test. Their search for emotional equilibrium is not made any easier by the environment. These women have experienced life in all its facets. For them, their family is not the sole center of their lives. They also like to make new friends.

The lifestyle desired by this group is life in the fast lane. These women love everything that is special, that is individual, that is precious, for example, "a trip to Italy for the grape harvest;" nor would they refuse an invitation from an "English lord for high tea." In their wardrobe, these perfume users place great emphasis on feminine elegance. While they do tend to have weight-problems, they would not be adverse to testing out all of the gourmet restaurants in the Rhône Valley.

The women in this group of perfume users are not, usually, among the very young.

The types of fragrances indicated for the individual personality groups are taken from the professional terminology used by the perfumer. They were named analogously in the Genealogy of Perfumes of Haarmann & Reimer. Genealogies of this type are structured around the basic raw materials contained in the perfumes and, consequently, the perfumes are classified by raw materials. For the perfumer who is interested in the structure of a perfume, this is certainly a good illustration.

However in the actual practice of a beauty consultant, this type of classification, based as it is upon raw materials, has very distinct limitations. The individual perfumes should be classified on the basis of the personality groups, so that the consultant can identify the type of personality behind a perfume. This is a pre-requisite for sound, helpful advice to a customer selecting a perfume.

Unfortunately, however, no psychological classification of perfumes and their users yet exists. An obvious method of course, would be to classify the perfumes included in a given fragrance-category, fragrance according to the types of perfume users we have identified. However, in comparing the fragrances allocated to the personality types with the perfume classification by raw materials, we find that there are variances. The study on individual perfumes and their user groups confirmed that this is the case.

The assessment of a perfume, by the nose or the personality of the user, is often quite different from that which the composition of the fragrance leads one to expect.

If we direct a particular user-group to one or the other of the established fragrance-categories we must remember that this can only be a general recommendation. Inevitably, there will be individuals (people and perfumes, both) that fail to conform to the general rule. Only a further analysis of the perfumes on the market will permit them to be precisely classified by perfume-user groups.

Fragrances Compared Culturally: the Influence of Culture and Life-Style, Geography and Climate

In the article entitled “Fragrance from the psychological side. The end of a myth?” the relationship to the area of fragrances was described in conjunction with an individual’s development from infant to adult.

In each phase of development, the individual learns new aspects of fragrance. During the course of this process, previous experiences are not lost; new ones are merely superimposed over the old.

In comparing the phases of development from infant to adult, it is possible to identify amazing parallels to the course of cultural development.

Cultural Development of Aesthetic Experience.

Since time immemorial, fragrances of all types have exerted a boundless fascination. Even the members of the gatherer and hunter cultures, in addition to their interest in fragrances and odors in conjunction with the satisfaction of their “primary needs" for food and sexuality, displayed an affinity for certain fragrances.

For the gatherer and hunter, and even for the people of the early high cultures, natural odorants were employed merely to mask unpleasant odors–an odor corrective–and as means of arousing the attraction and sexual interest of others. Moreover, fragrances were also employed in many cultures for paying homage to gods or godlike human beings. Conversely, even in early societies, odors that were perceived as being unpleasant were associated with evil spirits and demons. As late as the Middle Ages, the success of an exorcist's work was said to be evidenced by the smell of sulphur and other unpleasant odors as the devil left the mouth of the possessed person.

At these early stages of development, human beings considered themselves to be a part of an immutable world order and divine destiny. Under this regime, there was little leeway for experiencing and perceiving one's self as having an independent individuality. For the people of these cultural stages, their perception of themselves and access to their own personalities was, as opposed to today, much more indirect, less personal and linked to external characteristics.

There are pronounced parallels to this in the course of a child's development. Only as they grow older are children able to verbalize their wishes in the first-person form. Small children call themselves only by their first names, e.g.: "Charles is thirsty." When asked what they like about another child, children will cite external attributes, such as: A new toy, good at school, the strongest in the class, the best in the team, etc.

Just as it is not possible to speak of perfuming in the case of children, we cannot say that the people of early cultural stages perfumed themselves either, as we understand it today. The people of those cultures were able to experience their personalities only on an imparted and not very differentiated level. Their emotions were characterized by joy and anger, fear and satisfaction. It was not yet possible for the members of these cultures to perfume themselves for the purpose of rounding out their own personalities.

Only in the *Middle Ages* do initial signs of a new form of aesthetic experience become evident, finally achieving a dominating role at the beginning of modern times. More and more, individuals began to recognize themselves as independent beings, not ruled by some immutable, supreme system. Advances in agriculture, in particular, such as the introduction of the heavy plow; better utilization of animal labor through more effective harnessing techniques; and the transition to crop rotation; brought with them a new understanding of Nature and themselves. Increasingly, people became aware of the possibility of shaping their lives by actions of their own, depending less on divine dispensation. In this connection, the Church bemoans secularization. "Man and Nature have now become two different spheres, however man is getting the upper hand."

As a result of this gradual growth in new feelings of self-esteem, the individual began to reflect on the potential and versatility of his own personality. And, in doing so, he became aware of new moods and feelings. At the same time, the desire arose to find and assume responsibilities for one's own self.

A comparison with older children reveals similar interrelationships. They realize that their parents are not infallible, that they, too, can make mistakes. Older children increasingly question the order dictated by their parents. They demand to be treated as individuals and do not, for example, wish to be ranked equal with their sister or younger brother. And the older child also begins to become aware of its moods. Thus, for example, it can experience itself as being in a bad mood or very sensitive. In the Middle Ages, we can identify the initial signs of a form of aesthetic experience that is able, for example, to link fragrance creations, although initially still on a very primitive basis, with personal moods and feelings. Only at this cultural stage did human beings begin to develop into personalities that are able to underscore their characters by means of perfume.

It was only with the advent of *Modern Times*, i.e. after sloughing off the thinking that restricted and impeded the individual in the development and expansion of his or her personality, that the actual breakthrough came for this new form of aesthetic experience. Through the great intellectual advances of the Renaissance and humanism, there was a thoroughgoing and basic change in sociological, political and economic conditions. Renaissance and humanism are symbolic of a process of change, in which, through the interaction between a rediscovery of classical antiquity and a new, scientific approach to Nature, the individual began to shake off the bonds of the hierarchically structured Christian thinking of the Middle Ages. And in doing so, the individual became aware of his or her own intrinsic value as a temporal and historical personality.

Even as far back as the late Middle Ages, it was becoming increasingly undesirable to leave man and Nature to divine destiny. With the coming of modern times, Nature as a whole, came to be viewed as a system of matter that was subject to unchanging laws. Behind this change in the understanding of Nature can be seen a functional-rational method of thinking, bringing with it an explosion of scientific progress. Nature and man were investigated with unprecedented accuracy and thoroughness. Human beings discovered an interest in more and better knowledge. They changed their view of the world and became more self-assertive. There was a break with the classical view of the world, whose philosophy had, until Hegel, still been built around God. The philosopher Ludwig Feuerbach, on the contrary, viewed human beings as being God's creators and not vice versa. God exists only in our minds, said Feuerbach, in our imagination, not in reality and truth. Human beings project their ideal heavenward. With this loss of divine reference, it becomes necessary for human beings to reidentify themselves and their reason for living. In the search for self, they begin to understand themselves as being individual, independently thinking and feeling beings. With this newly discovered individuality, they are now able to develop a different qualitative relationship to the aesthetic. It is possible for them to identify their personalities–with their moods and feelings, in works of art and aesthetic products–and to relate themselves to them by means of their psyche.

However this new mode of experiencing, in turn, necessitates works of art and aesthetic products that permit this type of identification. Thus, for example, Medieval painting was not considered to offer a powerful enough statement for the Renaissance, and its effect upon mood and feelings was perceived as being not sufficiently mature and varied. Moreover, religious themes had lost their attraction and it was a view of a world oriented toward human beings and their history that was wanted.

The art of the Middle Ages is able to portray individuality only sporadically and imperfectly. In it, life has sharp contrasts, brilliant highlights and dark shadows, fresh and delicate complementary colors. The paintings of the Middle Ages do not yet show any transitions in portraying the emotional palette of Medieval man. The most violent colors are in glaring juxtaposition.

Traits of tenderness and gentleness can be found next to unthinking crudity. The wealth of sharply outlined heads, unique physiognomies, such as those in Renaissance portraits, cannot be found in the fine art of the Middle Ages. Renaissance painting nearly always portrays individuality, deliberately emphasing idiosyncrasy and originality, and even suggesting caprice or obstinacy. The Renaissance represents a complete break with the portrayal of Biblical themes. In the 17th Century, portrayals of idyllic landscapes were in full flower. Through poetic, well-judged lighting effects, painters now created "ideal landscapes," in which architectural backdrops, based on classical antiquity, impart a sense of temporal and historical awareness. The refinements of the mood were usually suggested by the hazy light of the morning or evening sun. For the aristocratic collector with a feeling and understanding for art, such pictures were intended to produce a solemn, uplifting atmosphere.

The difference between the Middle Ages and modern times also manifests itself in other forms of art, such as literature, poetry and music. Medieval music is still highly geared towards the Church and religion in its themes and orientation. In the Renaissance, music takes a more worldly trend just like painting. Music now appeals to the reason and feelings of the listener and demands involvement on these two planes.

Only at this cultural level is it possible for fragrances, like other aesthetic products and works of art, to become vehicles for self-actualization and self-realization.

Although the relationship to the world of odors and fragrances that had characterized previous stages of cultural development was not lost, it was greatly diminished in some respects. Through technical innovations and a change in our view of the world, fragrances gradually declined in importance for hygienic purposes through the development of modern plumbing and related facilities. Moreover, there was also a gradual change in nutritional habits. Food became less monotonous. Spices were no longer used as lavishly, but with more discrimination and in greater variety. Moreover, fragrances also declined in significance for hygienic purposes due to a continual decline in the number of apocrine sweat glands over the course of human development.

Even in the late Middle Ages, the use of fragrances in cults and rituals had diminished with the awakening of a functional-rational view of the world. Nevertheless, the significance of fragrances did not decline with the onset of the Renaissance. Quite to the contrary, it was only in the epochs that followed the Renaissance that fragrance came into its own. The discovery of the individual's own personality created a potential interest in a broad selection of fragrances.

Through an increasing number of technical innovations, primarily in the field of chemistry, and with the help of perfumers who had learned to express experiences and moods in terms of fragrance creations, it was possible to satisfy in ever greater measure the increasing need for self-actualization and self-realization. Henceforth, perfumery was highly esteemed in those societies that afforded the individual opportunities of personality-development and self-expression.

Moreover, we should be aware of the fact that not all cultural areas within a society are on the same level. Interculturally, there have always been differences. Regardless of their respective aesthetic preferences, not all members of a society have a comparable cultural level and need, as is reflected, for example, by the difference between soap users and perfume aesthetes. One fact can be recognized clearly: in all present-day societies, personality development is subject to change. Thus, today as in the future, comparable, yet specific personality traits will differ from society to society. And this fact should be taken into consideration in future fragrance creations.

Present-Day Societies and their Fragrance Need.

There are perfumes that sell very well in France, but not in Germany. There are countries, such as Brazil, for example, which, as opposed to its neighbors, such as Argentina and Bolivia, have their own typical fragrance need. There are perfumes that are popular in Europe and the United States, but must be modified for the Japanese market in order to be accepted there. What is the explanation for this?

An Intercultural Comparison of the three Decisive Aspects for an Individual's Fragrance Need.

In the article entitled "Fragrance from the psychological side. The end of a myth?" the question of why certain perfumes can have a fascinating effect on one person and not on others was examined. Studies show that people are fascinated by a perfume if this creation appeals to their personality in three areas:

The first area consists primarily of the emotions, i.e. emotional feelings and moods. In addition, the attitudes and associations of an individual to the type of fragrance also play a role.

The second area is that of an individual's desired lifestyle. Perfume users associate a given fragrance note or a specific perfume with a certain lifestyle. In doing so, of course, they are affected by the publicity statements on the product in question.

The third area consists of situations, occasions and socio-biographical data, such as age and sex; in addition, socio-cultural and climatic factors play a role.

The first area is of paramount significance in selection of a perfume. If a perfume does not appeal to perfume users in any of these three areas, they will reject it entirely.

If people are analyzed in these three respects, the reasons why perfumes exert a fascination on certain users can be seen clearly. The findings thus obtained result in seven broad perfume user groups. Since these studies were performed in the Federal Republic of Germany, the following questions arose from an intercultural standpoint:

- Can these seven perfume user groups also be identified in other countries?
- Which fragrance needs differ from country to country?

- Do the individual perfumes also have the same effect upon perfume users of other countries in these three areas, or are there variances?

In order to clarify these questions, countries that appeared to differ as greatly as possible from Germany were selected for a cultural-comparative study. Three South American countries were chosen: Argentina, Bolivia and Brazil.

First area: An intercultural comparison of emotional feelings and moods

Here, as in the German study, the perfume users from these three countries were investigated by means of appropriate psychological testing methods, in order to identify their personality features. In doing so, the following personality dimensions were of prime interest:

- Emotionally stable versus emotionally ambivalent; and

- Emotionally oriented toward the environment (extroverted) versus emotionally oriented inward (introverted),

The figure shows the two personality dimensions, divided into four quadrants.

Emotionally stable

Quadrant I	*Quadrant IV*
Introverted	Extroverted
mood trend	mood trend
Quadrant II	*Quadrant III*

Emotionally ambivalent

The studies showed that *German female perfume users* were found more frequently in Quadrants I and IV than in Quadrants II and III, there being a heavy concentration along the emotionally stable axis. These are perfume users who attempt to emotionally "get a grip on" themselves, to give the impression to those around them of being harmonious and well-balanced. In their thinking, they are realistic and purposeful. They are easily able to match their moods and feelings to the needs of the moment. It is obvious that, in the emotional area as well as in their attitudes, these perfume users reflect typical German educational standards and values, with which the individuals in question identify.

Corresponding to their personality features, a major portion of German perfume users lean toward the following two main types of fragrances: Aldehydic-floral, balsamic notes, which can extend all the way to Oriental notes; and secondly, toward chypre notes, which can extend all the way to fresh, floral notes.

As a result of their psychic attributes, on the other hand, *Brazilian women* are more frequently found in Quadrants II and III. These are people who can more aptly be termed as emotionally ambivalent. As opposed to the German, the Brazilian can be said to have less control over her emotions. She displays a variety of emotional feelings and moods, with the spectrum ranging from emotionally tranquil, introverted and withdrawn all the way to vivacious, active and cheerful.

The emotional spectrum of the Brazilian is also reflected in modern Brazilian literature, art and music. Of the three, the rich variety offered by Brazilian music is best known in the Northern Hemisphere. It ranges from the very popular "Saudade"–nostalgic, almost melancholy airs–to unbridled, euphoric moods, as expressed in hot Samba rhythms.

As opposed to German perfume users, the majority of the Brazilians favor the following two types of fragrances:

First of all the fougère family, which can extend to the Oriental notes, and, secondly, floral-fruity perfumes, that tend toward fresh-floral notes.

The fragrance need for fougère notes is an outstanding characteristic of the present-day fragrance market in Brazil. At the time of the study, no other country showed such great interest in variations on the fougère–lavender theme.

If the emotional feelings and moods of *Argentine* and *Bolivian female perfume users* are analyzed, a different phenomenon is encountered. As compared with German perfume users, both of these groups tend to adopt more extreme attitudes within the "Extrovert versus Introvert" personality-dimension.
A similar situation exists, although not so pronouncedly, in the "emotionally ambivalent" personality area. The same applied to the fragrance needs associated with specific emotional

personality features. Latin-American users with a preference for Oriental notes demanded this fragrance type much more definitely and emphatically than the German perfume users who inclined toward the same type of fragrance as a result of their emotional profiles. This extreme attitude showed again in the Latin-American woman's assessment of individual perfumes. Products that are most typical of their fragrance family are valued highly. In Japan, the situation was reversed.

Foreign perfumes were especially well received by female Japanese users if their respective fragrance characteristics had been toned down, i.e. made more harmonious and unobtrusive. The parallel to the Japanese character, and particularly their image of the ideal woman, is quite apparent. In a study of female Japanese perfume users, they would probably be found exactly opposite to Argentine and Bolivian users.

The intercultural comparative study performed in conjunction with the analysis of emotional feelings and moods shows the following quite clearly:

- The relationship between feelings and moods and corresponding fragrance preferences identified among female German perfume users also applies on an intercultural basis, i.e. internationally.

  Regardless of nationality, perfume users with an introverted mood trend have a fragrance need that tends toward Oriental notes in all countries.

- It was found that the same emotional personality features can occur in varying extremes in the individual cultures, which has repercussions on the respective fragrance character.

- National fragrance preferences are understandable if the predominant personality groups in each country are taken into consideration. Thus, the group of fragrance users in Brazil who incline toward the fougère notes as a result of their personality profiles is significantly greater than in other countries. Correspondingly, personality group comparisons provide a good indication why perfumes of a given fragrance type are more popular in one country than in another, in terms of their market share.

However, in order to explain why perfumes of the same fragrance type achieve varying degrees of acceptance in different countries, it is necessary to study the second and third fragrance-determining areas on an intercultural basis.

Second area:
An intercultural comparison of the desired lifestyle.

The studies in Germany show that female perfume users associate their emotional feelings and moods with a given, desired lifestyle. It is this lifestyle that is to be underscored by means of perfumes, and advertising messages about the product can influence the choice. Thus, it is possible to say that perfume users associate their perfume with the lifestyle that they are seeking.

A comparison with the three South American countries shows whether the lifestyle association with a perfume also applies on an intercultural basis.

The study, in fact, reveals that this is not the case. The lifestyle associated with individual perfumes in Germany is different from the lifestyle in the South American countries investigated, even in the case of comparable marketing concepts. The same perfumes are not considered by their users as a different fragrance type, but as having a different effect. For example, introverted German and South American perfume users have different associations with a certain classical Oriental fragrance (Genealogy of Perfumes, R 1925/11). For the vast majority of German users of this perfume, this creation today represents the ultimate in individual originality, which can be further emphasized by appropriate clothing. The wearer of this perfume differs clearly from other perfume users in her attitude toward life and her life-style. Her attitude toward life is "more progressive" and less tied to convention than that of the average woman. Moreover, her lifestyle is less oriented toward material things. On the contrary, she wants to experience herself differently and more intensively in her environment. In doing so, her perfume opens the door to herself and to a deep, mysterious and attractive, Oriental-exotic world, which represents her own ideal. In the final analysis, her life is devoted to seeking and discovering this other world in order to find her ideal in it. The search for her own self also manifests in her travel wishes, e.g. spending Christmas on a ship, on the

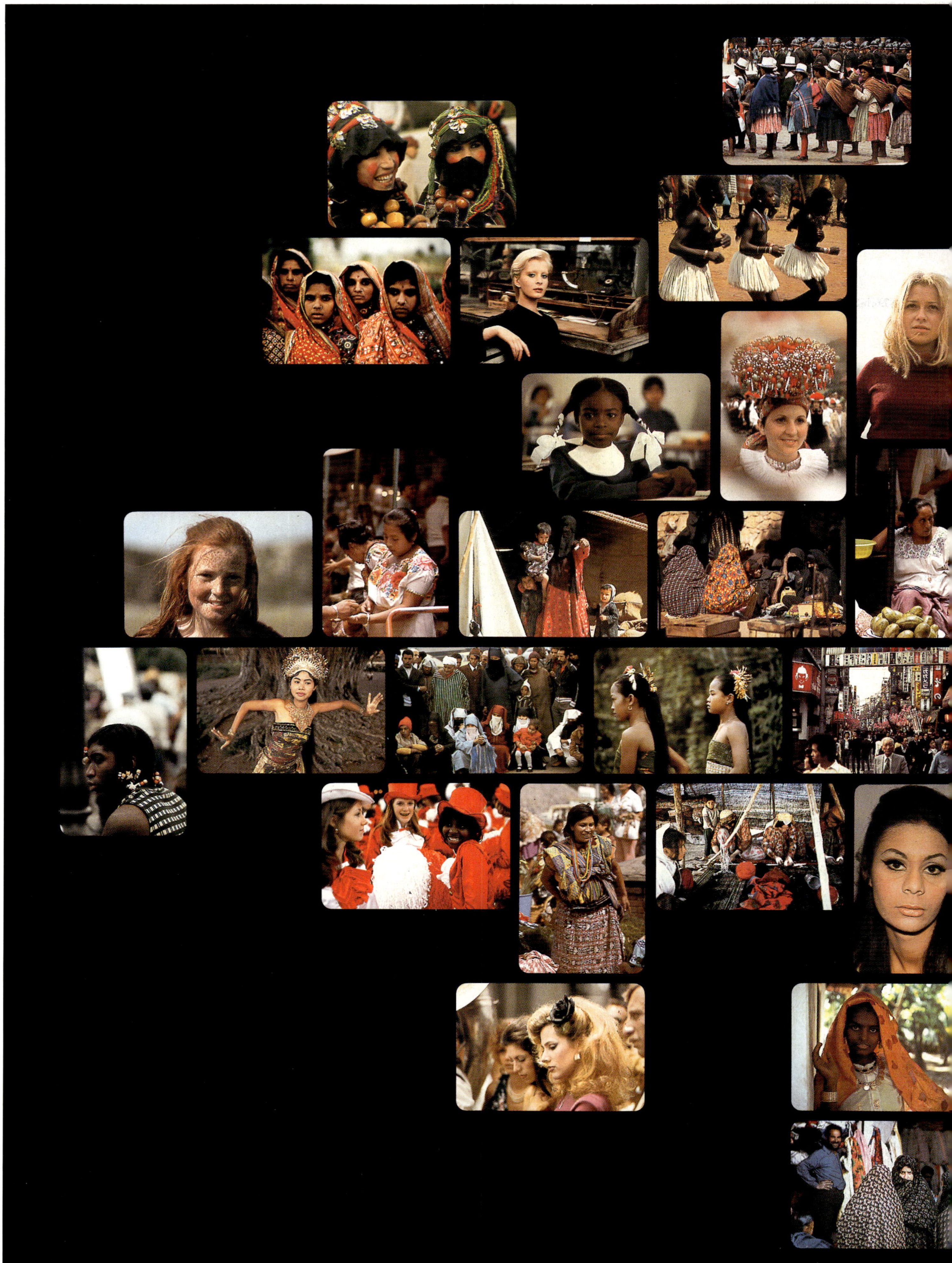

In forms of personality characteristics, people of all cultures are comparable. Does this extend to lifestyle, too?

Amazon amidst thousands of orchids and parrots; or strolling through the bazaars of the Orient.

The demands and lifestyle of these perfume users represent their response to a materialistic, achievement-oriented affluent society of the type that currently exists in the Federal Republic of Germany. Consequently, these attitudes and expectations are linked to such societies. Societies of an underdeveloped or developing nature offer less room for this attitude, and do not encourage it. People in such societies, which include Argentina, Bolivia and Brazil, form different expectations of life as a result of their indigenous cirumstances. In addition, differences in income structure between these societies and affluent societies are also reflected in the desired lifestyle of these persons. This is the reason why the lifestyle desired by the German users of the perfume cannot be identified in the three South American countries. Even if wearers of the perfume with comparable income levels and standards of living in these three countries are compared with German users, there is a difference. Argentine, Bolivian and Brazilian users of perfume whose standard of living is comparable to that of German perfume users are affected differently by the product: not in terms of their personality, but their lifestyle. For them, it is the French flair of the perfume that predominates. The perfume is, for them, a reminder and an expression of French culture and lifestyle, which they would like to make their own. In this connection, it is amazing to see how ancestors, relatives or travel can cause these individuals to relate to France. However this French trend is lost in relation to all perfume users in South America. For most of them, perfume (eau de toilette) is a "small luxury" that they do not wish to forego.

Generally speaking, it would appear to be questionable whether companies are well advised to take "lifestyle concepts" that are attuned to users in certain societies with a specific cultural background and a corresponding income structure, and transfer them directly to other societies.

As opposed to the first fragrance-determining area, the second area, the desired lifestyle, is more specific to the society in question. It is influenced by different cultural, historical and economic factors.

Third area:
Socio-biological data, cultural and climatic factors.

The influence of socio-biological data, such as age and sex, on perfume selection is subject to cultural considerations.

In Germany, for example, the users of Perfume 1925/11 (Genealogy of Perfumes) are mostly between 16 and 30 years of age, the majority being between 21 and 25. The relatively few over-30 users of this perfume view themselves as being younger than their age. This tendency could not be identified in the South American countries investigated. Argentine, Bolivian and Brazilian users of this perfume were more or less equally distributed throughout all age groups. Other perfumes produced similar results, although positions were reversed on occasion.

In the investigation of male or female preferences for certain odor notes and perfumes, there were also differences. Some fougère odors that are considered masculine in Germany were felt to be feminine by Brazilian women.

In South America, cultural and climatic factors have a major influence on the way in which individual perfumes are assessed. The degree of humidity in the atmosphere plays a significant role that should not be underestimated in the assessment of perfumes. In hot, dry air, the sense of smell is impaired to a greater or lesser degree as the nasal mucous membrane loses moisture, thus making it more difficult for the fragrance to be transmitted to the olfactory receptors. In tropical, hot-humid air, there is also an olfactory irregularity as a result of the intensive odor development in these climates.

In the temperate climate zones, it is also possible for humid air to enhance the olfactory sense when rain falls after a long, dry spell during the summer. However over-stimulation soon results in fatigue and a numbing of the sense of smell. An agreeable odor can suddenly become unpleasant. This is termed a swing phenomenon, i.e. at a given intensity, the quality impression of an odorant changes. Physiologically, this phenomenon can be explained by the trigeminus nerve, the free ends of which are located in the nasal mucous membrane. This nerve signals pain, warmth and pressure. Thus, there is an inverted U-shaped relationship between fragrance intensity and favorable assessment. As the intensity increases, an acceptable fragrance first becomes more and more pleasant. Excessive concentration, however, tips the scale, and the odor becomes more and more unpleasant. The point at which the fragrance swings from pleasant to unpleasant differs from individual to individual. Thus, it is possible for an individual to feel she is not wearing enough perfume whereas others will consider her overperfumed. The significance of humidity in connection with perfume assessment can be very clearly identified in several regions of South America. Test subjects in northeastern Brazil, a region with a hot, dry climate, criticized several famous perfumes as lacking both strength and tenacity. Test subjects living in warm, moist zones, e.g. in the Amazon region, soon assessed several branded perfumes as being flat and understated in an atmosphere already overcharged with odor.

Socio-cultural factors are also of decisive significance for the assessment of odors. In this connection, the article entitled "Fragrance from the psychological side. The end of a myth?" also deals with the significance of fragrance and perfume experiences made over the course of a child's development. In present-day societies, the interest in fragrances and odors is channelled by upbringing and is encouraged in only a few areas; usually, it is suppressed or denigrated as being insignificant. In the three South American countries investigated, education and convention appeared to have a less restrictive effect on individual fragrance experiences. In Argentina, Bolivia and, in particular, in Brazil, perfumes are received with much more enthusiasm than in Germany. Important in this connection, of course, is the fact that perfumes still hold a great attraction for many people in South America and are considered as something truly special. At the time the study was carried out, many well-known perfumes were only available in eau de toilette concentrations. In any event, the way in which South Americans relate to the fragrances is unlike that of German perfume users. In assessing a fragrance, all perfume users first decide whether they consider it to be pleasant or unpleasant. If it is held to be pleasant, the user can continue to explore the fragrance. If the perfume is judged unpleasant, most German perfume users have such a strong emotional aversion to it that it is no longer possible for them to make an objective assessment. The perfume is immediately "out of the running."

But most South American perfume users can still adopt an objective attitude toward a perfume, even if they basically reject it. They appear to have a generally more tolerant attitude.
In the study, other socio-cultural factors that are believed to influence perfume selection, such as different nutritional habits, body odor, different clothing, etc., were not confirmed as being "decisive in connection with perfume selection."

To summarize the results of the study:

The individual fragrance need of a person is very largely determined by her emotional feelings and moods, as well as by her desired lifestyle and socio-biological data (e.g. age, sex, geographic-climatic factors).

There is a total of seven personality groups, each with different fragrance needs.

Intercultural studies show that these seven personality groups, and their specific fragrance needs, can also be identified in other cultures. What changes from country to country is the size of the individual personality groups vis-a-vis the fragrance market. Of the three areas that determine fragrance need, and the choice of perfume, there is no change in the basic structure of the emotional feelings and moods, while the desired lifestyle and socio-biographical data are the products of each separate culture.

Perfume marketing:
Thinking in terms of the individual's personality.

Is there a difference between marketing a perfume and a brand of cigarettes? Not in terms of market research, experience suggests. Marketability is investigated in the same manner for both products. The researcher selects a potential target group, e.g. 16–25 year-olds with natural, uncomplicated personalities, for whom there is not much being offered in the marketplace. Appropriate products are created and tested to determine their marketability, i.e. 16–25 year-old women who project an uncomplicated, natural attitude are asked how they like the products. They are also questioned with respect to the products' design, packaging, etc. And this is basically how decisions are reached as to whether and how a new perfume or brand of cigarettes should be launched and how they should be packaged. But is decision-making based on conventional market research correct in the case of a perfume? Decidedly not!

Actually, we should have asked ourselves long ago why, in spite of costly market studies and high advertising expenditures, numerous perfumes appeal to consumers only briefly as being something new. Very often they are soon forgotten by the consumer and then pulled out of the market in order to make room for the next novelty. It is apparent that previous forms of market research experienced difficulty with perfumes. They treated perfumes and cigarettes in the same way, and thus failed to recognize what is special about perfumes. At the same time, they have stripped perfume of its real significance for the user. Perfumes are primarily an expression of moods and feelings. The perfume user reaches for her fragrance because she wants to produce moods and feelings that she senses within herself. And these feelings and moods, in turn, determine her fragrance need. Only if the fragrance need is satisfied by a perfume will the lifestyle and additional data associated with it play a role in the selection process. In order to ensure that a perfume tested in a market study will not by-pass the fragrance need, it is first necessary to identify the emotional personality profile of the target group in order to determine its fragrance need. In other words, it is necessary to think in terms of an individual's personality, in terms of her emotions, in order to find the type of fragrance that this individual is seeking.

In conventional market research, perfumes are placed before subjects whose ages and attitudes were known to the researcher. However there was no identification of their personalities in terms of their psychic characteristics. In fact, the market researcher knew nothing about the area that is most important for the selection of a perfume. Thus, for example, test subjects who are predominantly emotionally stable and introverted have been asked to pronounce upon a type of perfume that would have been perfect for emotionally ambivalent, extroverted perfume users, the target group that the perfume was in fact intended to reach. Is it then surprising that, as a result of this type of testing, perfumes that are well suited for a particular target group do poorly, or are not particularly well received, while, on the other hand, perfumes appear on the market although they have only earned half-hearted approval from the intended target-group? Half-hearted because, while the concept, the associated lifestyle, etc. were acceptable to the target group, the product did not completely satisfy their fragrance need.

The studies performed on fragrance selection clearly show the following:

Perfumes will, at best, be rated as "interesting" if they appeal only to the two fragrance determining areas of lifestyle and socio-biographical data, but fail in the first area of emotions.

What present-day market research does is to determine how a perfume will be received in the second and third areas. But the question of satisfying the fragrance need (area 1) of a target group is left to chance.

The Perfume Adviser – Everything You always wanted to Know about Perfumes

How do I find a perfume that is right for me?
Only by actually testing a fragrance on your own skin. In order to judge a perfume, you should live with it awhile. Only then will you know the times of day and the occasions that are right for that perfume and you. It pays to be a regular customer at a place where perfumes are sold, because then the sales people will be generous with samples. For the testing it is recommended that you use the less concentrated versions of a fragrance, such as the eau de toilette or cologne.

How does it happen that a perfume smells differently when you try it on in the store, and when you smell it again at home?
The impression of a perfume depends largely on the environment in which it is experienced. Perfumes are mostly tried in the heavily fragrance-laden atmosphere of the perfume department in a large store. These places are far different from the relatively neutral surroundings of a home, and as a result, the perfume will be perceived differently. Temperature and humidity differences are significant also. But the main reason is that every perfume has its own rate of evaporation and develops its true character only after a certain period of time.

What does the term "fragrance life" actually mean?
Perfume is a mixture of substances with different degrees of volatility. Fresh notes are predominantly light, highly volatile substances which characterize the first phase of the fragrance's life on the skin, the top note. Substances with less volatility produce the bouquet, the body, or in other words the middle note of the perfume. Substances with very low volatility and great tenacity produce the foundation.

A good perfume is built very carefully, with the objective that similar-smelling substances will be effective throughout all three phases of the fragrance life, making one continuous impression. At the moment of the first impact, the user is aware only of the initial odor–the top note. With good perfumes, there should be little difference between this and the succeeding phases. With not-so-good ones, the phases can be disturbingly different.

Does it always have to be the perfume, or can I also use the eau de toilette, say?
This is not only a matter of price. Some women love to use their favorite fragrances generously, and prefer to buy them in larger bottles. The following concentrations of fragrance (in an alcohol-water mix) are normally to be found on the market today:

	Concentration of Fragrance Essence
Extrait (or perfume)	15–30 %
Parfum de Toilette or Eau de Parfum	8–15 %
Eau de Toilette	4– 8 %
Eau-de-Cologne	3– 5 %
Splash Cologne	1– 3 %

Can you dilute perfume when it is too strong for you?
Yes. You can add alcohol. This should not be concentrated alcohol, however, but a solution of 90 % alcohol and 10 % distilled water.

How many different fragrances can one try without becoming numbed?
A variety of different perfume-types is less fatiguing for the sense of smell than a succession of closely-related odors. Consequently, one can say that 5–6 perfumes can be judged if they are all very different, but only 2 or 3 if they are of similar type. Further, your nose will tire more quickly of heavy fragrances than of light ones. So always begin with the lighter ones. And–fragrance samples should be applied as far apart from one another on the skin as possible.
Resist the attempts of any salesperson to cover your arms and wrists with too many scents.

Does a perfume smell differently on each person who wears it?
Yes. Everyone has her/his own individual smell. It depends on the condition of the skin, and on the life-style, particularly the diet. If a perfume smells good on another person, it's no guarantee that it will smell good on you.

Can the fragrance of a perfume change because of spicy food that its wearer might eat?
The fragrance of the perfume doesn't change, but the smell of the skin changes, because spicy dishes have an effect on the skin's exudations–and their odor. Consequently, there is also a change in the overall odor-effect, which is a combination of perfume and skin-odor.

Where and how should you apply perfume?
This is basically a question of the concentration of the fragrance. There's a good reason for the traditional drop-behind-the-ear. Perfume is most effective where the skin is especially warm and there is a good circulation of blood. Behind the ear, the wrist and the bend of the arm are excellent places for perfume. Freshly-washed hair is also a good fragrance-carrier, and so are clothes made of natural fabrics, and of course furs. Fragrance lasts particularly well on skin that is covered by clothing. Those who like to change their fragrances often should be careful about spraying them on clothes. The residual odor on a wool dress, for instance, might not be at all compatible with the fragrance you select the next time you wear it.

Why not use several fragrances at the same time?
Every perfume is a harmony in itself. If you use more than one at the same time, a disharmony generally results which destroys the characters of all of them. Even worse, it could cause an unpleasant impression.

Can you perfume your surroundings with a favorite perfume?
Yes. Some manufacturers offer room-spray versions of their perfumes. Economy tip: you can also spray a little perfume on your light bulbs. When the light goes on and the bulb gets warm, the fragrance is diffused into the atmosphere. Furthermore, perfume adheres well to upholstered furniture. But you can overdo room perfuming. Not everyone appreciates the same perfume and one can become irritated when a room, especially an office, is saturated with fragrance.

How can you give clothing and washing a pleasant odor?
There are wool detergents that contain perfume, so that you can give your woollens fragrance when you wash them. If you use a steam iron, you can put a little perfume into the water. Empty perfume bottles–stoppers out–give off their fragrance for a while in closets. If you like the idea, remember not to throw them away. But excessive perfuming of clothes and environment is to be avoided. If your scented clothes already make you smell like a perambulating perfume-sachet, you can't use fragrance as well!

When is a fragrance "obtrusive"?
Whenever a person has used too much, which can very easily happen with the most concentrated form of finished perfume, the extrait. But also, when the perfume type is too heavy and sweet. Over-perfuming can be particularly offensive in broad daylight, during the morning-hours. The rule is, the later the hour, the more concentrated the fragrance may be. But remember: when a fragrance is applied too often in the course of a day, the long-lasting components accumulate on the skin or the clothes and the result can be heavily obtrusive or jarring in its lack of harmony.

Politeness demands that one should use perfume carefully and intelligently. In a restaurant, for example, a cloud of perfume from a neighboring table can spoil a sensitive person's appetite.

Why is it that one is no longer aware of one's perfume after wearing it for a while?
Our sense of smell quickly becomes fatigued. After a period of smelling your own perfume, you become used to it and it seems to disappear. But to the by-stander, it can still seem very strong.

Are there rules on how to use which perfume when?
Yes. A perfume should never be allowed to dominate its surroundings. Less is always more appropriate than more. The more businesslike the occasion–let's say a business conference–the lighter, the more unobtrusive a perfume should be. During the day, the lighter fragrances such as the eaux de toilette or Eaux de Cologne are appropriate. For the evenings, the extraits, or more concentrated perfumes, are best. In the warm seasons, fresh, sparkling flowery notes are recommended. During the winter, the opposite. But fortunately, no strict set of rules exists. The

way a woman puts together a "fragrance wardrobe" to match her different moods and needs is wholly a matter of her own, personal taste.

Can a perfume influence mood?
Definitely. When an occasion is boring or particularly stuffy, or the mood in the office depressing, the stimulating fragrance of the right perfume can lighten the atmosphere surprisingly quickly. A pleasant fragrance can both stimulate and soothe, can bring a little happiness, relaxation, revitalization where none existed before.

Are there erotic perfumes?
Basically not, at least not in the sense of a direct-action aphrodisiac. We all know the pleasant scent of an attractive person of the opposite sex is definitely erotic, no matter which perfume he or she happens to be using. Heavy, warm fragrances give a feeling of closeness and intimacy. The lighter, fresher and more fruity a fragrance is, the less it is considered to be erotic.

Can you perfume yourself in such a way as to have your scent compatible with your companion's?
Perfumes do not exist in the so-called partner mode, which is all to the good. But there are quite a few men's fragrances that are used by women. Actually, feminine and masculine notes have been moving closer and closer together. Yet leather and some of the fougère notes still belong definitely in the masculine domain. If he should use a drop of her perfume from time to time, or she his Eau de Cologne, this would have nothing to do with the rapprochement of men's and women's fragrances–it would simply be a compliment to the favored fragrance of the other person.

Why is perfume so expensive?
One reason is the price of the raw materials. High-quality flower oils and animal fragrance substances have reached astronomical prices. Another reason is that perfume-production is a particularly expensive process. Years of work can go into the evolution of a new composition, and production is a long, multi-phase process. Another point: there are no economies of size, no family-packages in the haute couture perfume business. Manufacturers are not inclined to cheapen their products by selling them in larger sizes. There is also the fact that perfume can deteriorate if it is not used up fairly soon after the bottle is opened. The larger the bottle, and the air-space that grows as the contents are used, the greater the danger that the perfume will spoil.

Are natural fragrance substances better than artificial ones?
No. The perfumer differentiates between natural and synthetic or semi-synthetic substances. Natural fragrance substances are extracted in different ways (distillation is one) from vegetable raw materials such as blossoms, leaves, bark, wood, roots and the like; and from animal secretions. Semi-synthetics ("isolates") are separated from natural substances, and sometimes from processed products. Wholly-synthetic odorants originate one hundred per cent from the world of chemistry. All of these substances may be used in a single composition–they are of equal value to the perfumer. As to relative costs, there are extremely cheap essential oils from natural sources, and very expensive synthetic fragrance components. And vice versa.

What is a good perfume?
A perfume is good if it gives the wearer a sense of well-being, and if it is acceptable to people around her. There is the question of personal preference, but there are also objective criteria for the quality of a perfume. The fragrance must unfold, evenly and harmoniously, from the prelude of the top note, through the middle note, to the foundation without losing cohesion. In addition, the user should look for good tenacity, i.e. staying power. Part of the art of fragrance composition is the succesful building-in of fragrance retarders, the so-called *fixatives,* in such a way that the heavier, least-volatile and more stable substances hold back the lighter components. Fixatives can be natural fragrance substances such as resins, balms and animal secretions as well as macro-cyclic chemical odorants. On top of this, a good perfume has to be technically impeccable, meaning that it must be properly matured, carefully filtered, and dermatologically safe.

Is there any guidance for the choice of a perfume gift if you do not want to make a wrong choice in any circumstances?
This depends on the recipient. If she's of a mature age, the perfume in most cases would not be "Volcan d'Amour" or "Youth Dew." Conversely, if she's very young, something like "Indian Summer" would not be appropriate. The name of the perfume should be a help–it often expresses the perfume's character. Fragrances in a light, fresh,

flowery direction are acceptable to almost everyone. Heavier perfumes–those with more Oriental notes for example–are among the more risky presents because they are not as universally accepted. If you want to be as close to certain as possible, stay with the classics. A fragrance which has been a success for decades can't be as risky as the newest perfume on the market. In case of doubt, the donor must be guided by his own good taste.

Does perfume leave stains on clothing?
In most cases, no, because the vast majority of ingredients evaporates without leaving a trace. Nevertheless, it is not recommended to spray perfume onto a white silk blouse. Discolorations can occur when the perfume is too old, or dark, or has thickened. If stains occur, they can usually be removed by washing.

How can you tell if a perfume has spoiled?
A perfume has deteriorated when its color has drastically changed, when it thickens–and also, when you notice a sour, resinous smell on opening the bottle. When this occurs, some people are tempted to dilute the perfume with alcohol. But this will not help, the damage is done.

How long does perfume keep?
Around six months, usually longer, if it is stored correctly. The ideal storage conditions are in darkness, at room temperature. A new bottle of perfume will keep, unopened, for about a year. Once the bottle has been opened, use the contents as quickly as possible. The idea of keeping perfume in the refrigerator is of doubtful value–the perfume traces on the outside of the bottle could affect the food.

Can I make perfume-mixtures of my own?
No. Each perfume is a complex, balanced combination of many different ingredients, some of which occur in the tiniest of amounts. It is simply not possible for the layman to obtain these substances. A do-it-yourself perfume kit does exist, however. It consists of finished compositions which you can mix according to your choice. The results can be pleasing, but the whole idea is more of a gimmick than a real perfume-making kit.

How many substances does a perfume contain?
There is no rule. Perfumes can be composed of sixty, seventy, of hundreds or ingredients–the sky is the limit. But this is not to say that a large number of ingredients automatically signifies high quality.

Should a perfume be dabbed on or sprayed?
This is a question both of economy and comfort. When you dab perfume on, precious drops can be wasted. Aerosol spraying facilitates a fine, even application.

The so-called atomizers, or pump-sprays, also deliver the perfume as a fine spray, but not continuously, as in the case of aerosols. The lighter fragrances, such as the eaux de toilette are being offered more and more frequently with a choice of either type of spray package.

How many millilitres equal one ounce?
One ounce equals 29.57 millilitres.

Author Index

Our Sincere Thanks to

Perfume:
from the Concentrated Composition to
International Brand
Julia Müller

Fragrance from the Psychological Side.
The End of a Myth?
Joachim Mensing, Christa Beck

Fragrances Compared Culturally:
the Influence of Culture and Life-Style,
Geography and Climate
Joachim Mensing, Christa Beck

The Perfume Adviser –
Everything You always wanted to
Know about Perfumes
Julia Müller, Egon Oelkers

Messrs. Pierre Chauvet S.A., Seillans, France, for permission to photograph on its premises and for providing editorial material and graphics concerning essential oils and manufacturing processes.
(Pages 74, 75, 94).

Perfume Museum Regia, Barcelona, Spain, for the providing of antique perfume containers (Pages 20, 21, 22, 23, 25, 27, 28, 29, 30, 31, 32, 34/35, 36, 37).

The Firm of Jacomo, Le Havre, France, and the Firm of Juvena, Baden-Baden, West Germany, for permission to photograph on their premises.

Mr. Hans-Hermann Rief, Worpswede, West Germany, for the providing of antique perfume containers (Pages 62/63).

Parfümerie Douglas, Göttingen, West Germany,
Parfümerie Douglas, Munich, West Germany,
Koron Parfümerie direkt am Dom, Munich, West Germany,
for the providing of display materials and perfume flacons.

Association of German Druggists, Cologne, West Germany, for its help to formulate the questions in the article:
"The Perfume Adviser. Everything You always wanted to Know about Perfumes."

Picture Reference

Explanations:
l. = left, r. = right
t. = top, m. = middle
b. = bottom

Photography

Manfred Blume, Holzminden, West Germany
(Pages 148/149)

Pierre Chauvet S.A., Seillans, France
(Pages 92, 95 r., 96 l.t.)

Peter Gauditz, Hannover, West Germany (Page 87)

Gruner + Jahr Fotoservice, Hamburg, West Germany
(R. Eisele, P. Eising, M. Friedel, Dr. J. Gebhardt,
R. Hayo, R. Joedecke, P. Knapp, G. Mangold,
F. Schieler, J. Schmitt, P. Thomson, H.-R. Uthoff)
(Pages 148/149)

H&R-Archives, Holzminden, West Germany
(Pages 46, 108, 109, 112, 113, 114, 115, 148/149)

Udo M. Klumpp, Munich, West Germany
(Pages 20 b., 21, 22, 23, 25 r.t., r.b., 27, 28, 29, 30, 31,
32, 33, 34/35, 36, 37, 100 l.)

Siegbert Kercher, Hannover, West Germany (Page 24)

Dieter Klemme, Holzminden, West Germany
(Pages 91 l.b., 93 r.b., 95 l., 102 l.t.)

Studio Kussmann, Frankfurt a/M, West Germany
(Pages 38, 39, 40, 41, 42, 43)

Siegfried Rogas, Holzminden, West Germany
(Pages 88, 89 t., 104 r.b., 107 r.)

Erich Schuster, Opladen, West Germany
(Pages 20 t., 25 r.m., 26, 45, 89 l.b., r.b., 90, 91 l.t.,
r.t., r.b., 100 r., 101 r.b., 103, 104 t., l.b., 105, 106)

A. + P. Schudel, Zurich, Switzerland
(Cover, Pages 116, 117, 118, 119, 120, 121, 122, 123,
124, 125, 126)

Reinhart Wolf, Hamburg, West Germany
(Pages 97, 98, 99, 101 t., l.b.)

Illustrations

Medical Service, Munich, West Germany
(Pages 50, 51, 52, 53, 54, 56)

Fred Niemüller, Ronnenberg-Benthe, West Germany
(Pages 62/63, 74, 75, 80, 81, 82, 83, 84/85, 94,
136/137, 138, 139, 140, 141)

Dieter Wullkopf, Hamburg, West Germany
(Pages 11, 12, 13, 14, 15, 16)